How Culture Makes
Us Human

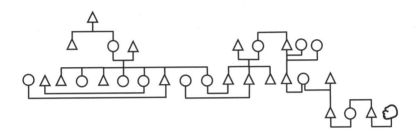

How Culture Makes Us Human

Key Questions in Anthropology: Little Books on Big Ideas

Series Editor: H. Russell Bernard

Key Questions in Anthropology are small books on large topics. Each of the distinguished authors summarizes one of the key debates in the field briefly, comprehensively, and in a style accessible to college undergraduates. Anthropology's enduring questions and perennial debates are addressed here in a fashion that is both authoritative and conducive to fostering class debate, research, and writing.

Proposals for books in the series should be addressed to ufruss@ufl.edu. Series Editor H. Russell Bernard (emeritus, University of Florida) has been editor of the journals *American Anthropologist, Human Organization,* and *Field Methods,* and of the series Frontiers of Anthropology. He is author of the leading textbook on field methods and has published extensively in cultural, applied, and linguistic anthropology. He is recipient of the prestigious AAA Franz Boas Award.

Series Titles

Archaeology Matters: Action Archaeology in the Modern World
    Jeremy A. Sabloff
The Origin of Cultures: How Individual Choices Make Cultures Change
    W. Penn Handwerker
How Culture Makes Us Human: Primate Social Evolution and the Formation of Human Societies
    Dwight W. Read

How Culture Makes Us Human

Primate Social Evolution and the Formation of Human Societies

Dwight W. Read

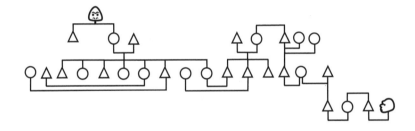

Walnut Creek, CA

Left Coast Press, Inc. is committed to preserving ancient forests and natural resources. We elected to print this title on 30% post consumer recycled paper, processed chlorine free. As a result, for this printing, we have saved:

2 Trees (40' tall and 6-8" diameter)
1 Million BTUs of Total Energy
213 Pounds of Greenhouse Gases
960 Gallons of Wastewater
61 Pounds of Solid Waste

Left Coast Press, Inc. made this paper choice because our printer, Thomson-Shore, Inc., is a member of Green Press Initiative, a nonprofit program dedicated to supporting authors, publishers, and suppliers in their efforts to reduce their use of fiber obtained from endangered forests.

For more information, visit www.greenpressinitiative.org

Environmental impact estimates were made using the Environmental Defense Paper Calculator. For more information visit: www.papercalculator.org.

Left Coast Press, Inc.
1630 North Main Street, #400
Walnut Creek, CA 94596
www.LCoastPress.com

ISBN 978-1-59874-588-7 hardcover
ISBN 978-1-59874-589-4 paperback

Library of Congress Cataloging-in-Publication Data

Read, Dwight W., 1943-
 How culture makes us human : primate social evolution and the formation of human societies / Dwight W. Read.
    p. cm.—(Key questions in anthropology : little books on big ideas)
 Includes bibliographical references.
 ISBN 978-1-59874-588-7 (hardcover : alk. paper)—ISBN 978-1-59874-589-4 (pbk. : alk. paper)
 1. Primates—Evolution. 2. Human evolution. 3. Social evolution. I. Title.
 QL737.P9R315 2011
 616'.02738—dc23

                              2011028973

Printed in the United States of America

    The paper used in this publication meets the minimum requirements of American National Standard for Information Sciences—Permanence of Paper for Printed Library Materials, ANSI/NISO Z39.48–1992.

Cover design by Jane Burton

Contents

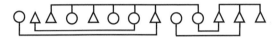

I am deeply indebted to my wife and colleague, Fadwa,
for her constant and heart-felt encouragement.
Without her suggestion that I should organize,
in book format, the ideas I had been presenting
in talks at academic meetings,
this book would not have been written.

Preface

This book has had a long gestation and draws upon my background as both a mathematician and an anthropologist. Mathematics highlights the power of formal modes of reasoning as a way to work out the logical consequences of a series of premises. Anthropology makes evident the importance of culture in human societies, not as a determinant of behavior, but by providing a shared framework or context within which we interact in a mutually understandable manner in accordance with the organization, structure and mode of adaptation constituting the social and cultural system of which we are a part. My long term interest has been to bring these two forms of knowledge together so as to better comprehend the nature, form, evolutionary history and possible future direction of human societies.

Themes in this book have already been presented, in preliminary form, in a number of venues. A key part of my argument relates to the implications derived from the increased complexity of the social field arising from expanded individuality of behavior during the evolutionary history of ourselves and of other primates. This increased complexity has led to major changes in social systems, regardless of whether it is the social field of an individual localized within a troop of Old World monkeys (chapter 2), a community of chimpanzees (chapter 4) or a group of hunter-gatherers (chapter 3).

While the extensive behavioral individuality of the great apes has long been recognized, the implications of increased individuality for change in primate social systems based on face-to-face interaction have not been adequately considered. A social field composed of dyads and not just individuals increases exponentially in complexity with increased individuality of behavior. However, the implications of an exponential increase in social complexity as a driver for change in the form of social organization in primates from monkeys to the great apes and then humans has not been given sufficient attention. In a talk I gave at the First Annual Conference on Multi-Agent Modeling in the Social Sciences (held at Lake Arrowhead, California, in 2002 and organized by the UCLA Center for Human Complex Systems) titled "The Emergence of Order from Disorder as a Form of Self-Organization," I discussed the implications that increased behavioral individuality has for social complexity and how this relates to the transition from non-human primate to human forms of social organization. The talk was subsequently published in the journal *Computational & Mathematical Organization Theory* (Read 2004).

I explored the role of culture for change in the organization of human societies, as one goes from band level to state systems, in an invited Plenary Talk for the Second International Conference on Complex Systems (Nashua, New Hampshire, held under the auspices of the New England Complex Systems Institute in 2002). The talk, titled "The Role of Culture in the Emergence of Complex Societies," explored the interplay between culture in the form of a conceptual system of kinship relations and change in social organization, a theme taken up in chapter 6.

The increased complexity of the social field arising from increased individuality in behavior relates more broadly to the evolutionary trajectory going from systems of social organization based on face-to-face interaction to the relational basis for interaction that characterizes human societies. This transition involved more than just the elaboration of pre-adaptations and/or cognitive and behavioral capacities that were already present in an ancestral primate species, but centered especially on the evolution of cultural systems, and hence related directly to what we mean by cultural evolution. This theme was presented as a two-part seminar titled "On the Origin and Evolution of Culture" given at the Center for the Study of Evolution and the Origin of Life (UCLA) in 2003 and 2004.

Subsequently, I was invited to participate in a multi-year research project titled "Information Society as a Complex System" (ISCOM) funded through the European Union's Directorate for Information Science and Technology and coordinated by the University of Modena and Reggio Emilia in Italy. This project provided me with the opportunity to develop further, through a series of presentations in project meetings, the idea that the transition from non-human primate to hunter-gatherer forms of social organization was not simply one of elaboration and expansion on nascent capabilities already present in our primate ancestry, but entailed a fundamental innovation in the mode of evolution, changing from Darwinian individualistic and population based evolution to evolution at the level of societal organization. The primary project directors, David Lane at the University of Modena and Reggio Emilia and Sander van der Leeuw at the Arizona State University, provided critical feedback and helped identify areas where the argument needed further elaboration. Our discussions led to three co-authored chapters in the book *Complexity Perspectives in Innovation and Social Change* that came out of the ISCOM project. A key idea developed in these chapters is that "human social change cannot be described in Darwinian terms, because something new has appeared, i.e., the fact that human societies are inherently responsible for their own innovation" (Lane et al. 2009: 4). This innovation, developed and discussed in detail in the chapter "The Innovation Innovation," for which I was the senior author, provides the conceptual foundation for this book. I have elaborated further on this theme in the paper I was asked to present at the conference "Social Brain, Distributed Mind" held under the auspices of the British Academy in 2008 and published with the title "From Experiential-based to Relational-based Forms of Social Organization: A Major Transition in the Evolution of *Homo sapiens*" (Read 2010b).

In this book I begin with the premise that human societies are, in fundamental ways, unlike any other animal society, yet what makes for that uniqueness cannot be measured in a simple manner. We are all familiar with the various indices of what supposedly makes humans unique that have fallen to the wayside as we increase our understanding of the richness of animal societies, especially our primate relatives. At one time we were said to be the only species that made and used tools on a regular basis, but we now know that chimpanzees in the Old World and capuchin monkeys in the New World both make and use tools in a regular and adept manner. Or, we were said to be uniquely a species with a full-fledged language ability, but it is increasingly evident that language-like abilities are not unique to us and that other species have sophisticated systems of communication that enable complex forms of social behavior and organization.

We can always tweak the behavioral distinctions so as to make it appear that we are a unique species. Rather than tool makers and users, we are now said to be the only species that uses tools to make tools or we are the only species that has an elaborated system of communication predicated upon integration of syntactic and semantic distinctions. This tweaking, though, only underscores the fact that our uniqueness does not lie in any one behavior, but more broadly in the cognitive and conceptual capacities we bring to bear in the formation of our specific forms of social behavior (Penn, Holyoak, and Povinelli 2008). Social insects have complex and marvelous forms of social behavior and organization with a remarkable degree of parallels between their social behaviors and forms of organization and what we find in human societies, yet they differ from us, fundamentally, through having social systems dependent upon explicit, genetically determined behaviors. Our societies are not based on genetically prescribed behaviors, but on our ability to build creatively upon our genetic endowment in ways that are not predictable simply by knowing what constitutes that endowment. We can see this in the way human societies have evolved and elaborated themselves since the Upper Paleolithic without first requiring additional genetic changes and in ways that would have been unimaginable to our ancestors of that time.

In our evolution as hominins, we have undergone two major phases. The first extends from around 8 million years ago when we last shared a common ancestor with a non-human primate species to the Upper Paleolithic and is characterized by the evolutionary development of the morphology and cognitive abilities of our species, *Homo sapiens*. These changes were in place by the time of the Upper Paleolithic, though aspects of them had appeared previously. The overall degree of change in our cognitive abilities is reflected in the increase in our short term working memory (STWM). STWM has gone from 2 ± 1 for our common ancestor with the chimpanzees (Read 2008c) as implied by 25 percent of the chimpanzees at Bossou, Guinea, in West Africa not being able to learn to integrate together three objects (an anvil, nut and hammer stone) as a way to crack nuts, to the 7 ± 2 characterization of human STWM made famous by George A. Miller's (1956) article on the size of short term memory in modern humans.

The size of short term working memory relates to the number of different concepts, ideas or dimensions we can integrate together simultaneously in our thinking. Thus it measures the degree of integrative complexity that we can activate in our interaction with the physical and social environments in which we are embedded and which we also construct for ourselves. We can get a rough sense of the degree of conceptual complexity introduced during hominin evolution of STWM leading to the Upper Paleolithic through the increase in the conceptual dimensionality of artifacts made by our ancestors. Tool complexity has gone from Oldowan choppers made by *Homo habilis* 2.5 million years ago with a conceptual dimensionality of three (the working edge of the chopper that is being made, the angle of percussion needed for control-ling flake removal to make the edge, and the controlled use of a hammer stone to remove flakes must be conceptually integrated), to a dimen-sionality of around seven for the Levallois flakes and prismatic blades made during the Middle Paleolithic and the soft hammer, prismatic blade technology that appears in the Upper Paleolithic. Production of new kinds of tools using these technologies involves geometric con-ceptualization in three dimensions and recursively employs, for their production, an algorithm for flake (Levallois technique) or blade (soft hammer, prismatic blade technique) removal (Read and van der Leeuw 2008, and references therein). Tool making in the Upper Paleolithic in Europe, North Africa and western Asia elaborated extensively on blade technology and involved wide-spread use of multi-component, hafted blade tools (Bar-Yosef and Kuhn 1999) whose conceptualization required a STWM comparable to that of *Homo sapiens* today.

These cognitive abilities associated with our species in the Upper Paleolithic have been described under the rubric "enhanced working memory," due to our ancestors having an expanded role for the execu-tive functioning of working memory (Coolidge and Wynn 2009), with the executive function of working memory responsible for attention and decision making (Baddeley 2001). Altogether, the cognitive changes in place by the time of the Upper Paleolithic made for a new kind of adaptation—an "organizational rubicon" (Leaf and Read forthcoming: 41)—based on new organizational forms constructed using linguistic and cultural capacities made possible through the evolutionary changes that had already taken place in the cognitive abilities of our ancestors.

These organizational changes introduced a second phase in our evolution characterized by a shift to direct change in the form and mode of social organization rather than indirect change by individ-ual traits leading, through emergence, to group level properties. This shift can be seen in the development and subsequent evolution of the forms of social organization characterizing human societies as part of our increasingly complex modes of resource procurement as we have gone from hunter-gatherer to agriculture based societies. As hunter-gatherers, we lived in societies of around 500 persons organized as interconnected bands of 30–35 persons (chapter 6). From this we have gone to large-scale state and even multi-state forms of social and politi-cal organization and structure without any apparent major changes in cognitive abilities. Unlike the first phase where we find a close paral-lel between changes in cognitive abilities as measured by, for example,

STWM and the complexity of the mode of adaptation as indicated by the artifacts used in that adaptation (Read and van der Leeuw 2008), the second phase builds on the cognitive abilities already in place by the time of the Upper Paleolithic and elaborates on changes that were possible with already existing cognitive abilities. The second phase qualitatively differs from the first by the way changes at the group level affecting the form and mode of social organization had become directly part of a group's adaptation, rather than arising indirectly through changes at the individual level (chapter 6).

We can see evidence of change at the group level in the development and central importance of kinship systems in the formation of human societies (chapter 5). Kinship in human societies structures the initial social domain in which we are embedded from birth and both defines and provides conceptual ways for individuals to identify the kin relationship they have to one another. We can usefully distinguish between kinship in this conceptual sense and biological kinship arising from reproduction by referring to the former as cultural kinship. Although cultural kinship is still ultimately based on the biological facts of reproduction, it transcends its biological roots through forming conceptually expressed systems of kinship relations that need not parallel biological kinship relations (Read 2001). Cultural kinship became the innovative framework (Read, Lane, and van der Leeuw 2009) through which new systems of social organization are defined and expressed. Systems of mutually understood kinship relations provide the social boundary for those among whom social interaction takes place and are the basis for an expanded social field that can include individuals not currently residing together (chapter 5). By the time of the Upper Paleolithic, the social field of interacting individuals is no longer limited, as it is with non-human primates, to those with whom face-to-face interaction on a daily basis is taking place (Gamble 1998, 2010). Instead, the social boundary becomes defined conceptually. As a consequence, human social systems have a cultural boundary rather than the biological boundary that distinguishes sexually reproducing species (chapter 6).

It is here where we find a fundamental change that distinguishes human social systems from the kinds of systems of social organization we find in other species. Human social systems are not an elaboration of either genetic systems or genetic systems coupled with face-to-face interaction as the basis for social organization, but have a new form of organization in which the social domain is determined through a conceptual system of social relations expressed both genealogically through parent-child relations as they are culturally understood and linguistically through a kinship terminology (chapter 6). The kinship terminology—the linguistic terms used to express the kin relations one person has to others, such as the English kin terms mother, father, brother, sister, uncle, aunt, father-in-law, mother-in-law, and so on—is not simply a collection of terms for categorizing already existing kin relations, but is a conceptual system that expresses what are the kin relations as they are understood in a particular society and how kin relations may be computed among the individuals sharing the same kinship terminology (Read 2007) to form what is sometimes called a kinship based society. Associated with kinship relations are expected patterns of behavior characterized by Meyer Fortes (1969) through his axiom of prescriptive altruism (or amity)—namely that kin are expected to be mutually supportive even if only by virtue of being recognized as kin through a conceptual system of kin relations, independent of biological kin distance (Palmer and Steadman 1997). For prescriptive altruism to operate throughout the social field, it is necessary that the social boundary be expressed through those who are mutual kin to one another (chapter 6), hence the social field will be made up of those who mutually share, through enculturation, the same conceptual system of kinship relations and expected, supportive behavior (although such behavior is not always realized) of one kin to another.

The odyssey from the Old World monkeys to the great apes and then to the development of our unique forms of social organization is, then, the overall theme of this book. The odyssey begins, as it must, with our biological roots as a primate and with change in social organization initially occurring through Darwinian evolution. The challenge, when coming forward to human societies as we know them, has been to connect our Darwinian beginnings to the current complexity of human social systems in which Darwinian evolution, with its focus on individual traits in the context of a population of interbreeding individuals, has been transformed into a new mode of evolution with change at the level of societal organization. It is the functionality of systems of organization, rather than the functionality of individual traits, that is critical to the evolutionary success of human societies. Ancestral hunter-gatherer societies developed cultural means for the expression and continuity over generations of societal practices from whose functionality individuals and families benefit. Through enculturation, individuals take on the properties, structure and features that are part of the cultural milieu that frames the way individuals and groups of individuals interact. More than a century ago, Edward B. Tylor (1924[1871]: 1) referred to culture as "that complex whole," a characterization that still stands today as a way to identify what is different about human societies in comparison to the societies of other social mammals. It is not an extra-somatic means of information transmission that is crucial to what constitutes culture, but rather that culture refers to conceptual systems such as the kinship systems central to the formation of human societies. The origin of kinship systems as the basis for social organization, hence as a conceptual system of relations, encompasses a transition to new forms of organization subject to change by the individuals embedded within those systems of organization. It is this capacity for self-modification that makes human societies unique.

CHAPTER 1

Introduction

Probably one of the most interesting—and challenging—stories in the evolution of our species is the transition from our shared ancestry with other primates to human societies as we know them today. What makes us like other primates and what makes us different? These are the evolutionary themes worked out during this crucial transition. These themes help define what distinguishes our species from other primate species.

Two main ideas have been used to characterize this transition. One idea sees our species, though distinct from other primate species in specifiable ways, as essentially the culmination of trends already present in primate evolution. Accordingly, we can use our understanding of those trends to comprehend the characteristics and properties of our species: we can make inferences and analogies between humans and primates in the same way we do between non-human primate species. The other idea sees our species as having undergone changes that have restructured us in such a way that we have, through culture, what the French anthropologist Claude Lévi-Strauss called "the advent of a new order" (1969[1949]: 25). In other words, the internal dynamics of how we are socially structured and organized depend upon processes unlike those we find in other primate species.

According to the first viewpoint, the difference between us and other primates is primarily quantitative, not qualitative. Further, although the way in which human societies are culturally based has been used to argue that we are qualitatively distinct, from this perspective it is argued that a boundary based on culture is neither sharp nor clear cut. We find the rudiments of culture, it is claimed, in other primate species and according to some, even in non-mammalian species as different from us as crows and magpies. While the claim that the rudiments of culture are found in avian species may beg the question

Figure 1.1 Hand-clasping while grooming. *Photo by Vernon Reynolds, used by permission of publisher.*

of what we mean by culture and what are its cognitive prerequisites, it is difficult to discount behaviors among primates closely related to us, such as the common chimpanzee, that appear to have all the earmarks of what we mean by culture, at least in a rudimentary sense.

Rudiments of Culture

The chimpanzees in Mahale Mountains National Park, Tanzania, have (to us) a peculiar way to groom that satisfies the criteria often used to distinguish what is cultural. Two grooming chimpanzees raise one arm each and clasp their raised hands (see Figure 1.1). While holding their arms upward in this manner, they groom each other with their free hands. Only chimpanzees in this region groom in this manner, indicating that it is not a trait passed through genetic reproduction, but rather transmitted directly from one individual to another in a social context. It is "learned (rather than instinctive), social (rather than solitary), normative (rather than plastic), and collective (rather than idiosyncratic)" (McGrew 2003a: 433). Were this a human trait, we would easily accept it as a cultural behavior.

If, as this example suggests, chimpanzees are capable of expressing rudimentary forms of culture, then, according to this line of thought, the expression of what we call culture in human societies need not be anything more than an extension and elaboration of capacities we already find in chimpanzees. However, our species arose from a last common ancestor with the chimpanzees, not from modern chimpanzees, and so to complete the argument, we need to know if the last common ancestor to the chimpanzees and ourselves exhibited such behaviors.[1] Whether it did is unknown, but there is no reason to rule out the possibility that an ancestral species for us and the chimpanzees had the cognitive capacity to develop rudimentary forms of culture from which, some argue, our more elaborated forms of culture evolved through Darwinian evolution.

Similarly, if we examine the forms of social organization we find in non-human primate societies, we appear to find the beginnings of some of the organizational aspects that are central to the organization of many human societies. For a number of Old World (OW) monkey species, the troops into which a species is typically divided have a form of social organization based on matrilines composed of females connected through biological mother-biological daughter links. The central players for a monkey troop organized in this manner would be a few elderly females, the biological daughters of these females, and their daughters' biological daughters. Males born into the troop leave at adolescence and join other troops when they begin mating. Humans also have matrilineal societies based on lineages; that is, a social unit made up of individuals whose ancestry through mothers traces back to a person recognized as the focal, or founding, ancestress for the lineage. Regarding membership, a matrilineage is built around a few elderly women, their daughters and their daughters' daughters, along with the sons (but not the sons' children) of women in the lineage. For lineage-based societies, marriage is typically exogamous—there is a cultural rule that specifies that one must marry someone outside of one's natal lineage. A male born into a lineage will marry a female from a different lineage, and though there is variation across matrilineal societies, he may leave his natal lineage upon marriage and join the lineage of his bride, much like, it is argued, a sexually mature male leaving his natal primate troop to mate with females in another troop. It is easy to find other examples of what appear to be similar behaviors and forms of organization in human societies and non-human primate societies, such as the pair-bonded gibbons, whose behavior matches the monogamous marriage rules of human societies.

From an evolutionary perspective, however, similarity is not enough to establish common origin. Evolutionary biology distinguishes between the case in which a particular trait is heritable and found in two populations because both populations are descended from a single, ancestral population that has the trait, and the case of analogy in which a particular trait exists in the two populations because of convergent evolution, perhaps because each population has adapted to similar conditions. Analogous traits are often functionally similar but arise in different populations for reasons unrelated to phylogenetic connection. A similar caution applies to the matrilines considered above. Does the similarity in the structure of a primate troop built around female matrilines and the structure of a matrilineage, also built around female matrilines, arise from common inheritance from an ancestral species? In this case the answer is "no" for we know that there is no direct connection between the troop structure of the OW monkeys and the presence of matrilineages and matrilineal societies. How do we know? Because the matriline structure of the OW monkeys is not carried forward to any of the great apes and hence most likely was not part of the social organization for a last common ancestor of our species and the chimpanzees.

Cognition and Behavior

Even with homologous traits, and especially traits relating to the cognitive abilities of ourselves and our primate ancestors, it is not sufficient to just show continuity. How the traits are implemented—the phenotype in comparison to the genotype—must be taken into consideration. For cognition, the phenotype is quite plastic as the biological specification of many aspects of cognitive functioning has to do with capacities and not specific abilities. When we refer to intelligence, we are talking about a wide range of abilities that can be expressed to a varying degree even with the same underlying cognitive capacities. The difficulty in resolving the nature-nurture debate for cognitive performance simply underscores the fact that in the cognitive domain there is not a tight relationship between genotypic specification and phenotypic expression. At the neurological level, similar brain architecture between ourselves and the chimpanzees suggests continuity in cognitive capacity, yet there are qualitative differences between human and ape brains (Rilling 2005). Even between the closely related common chimpanzee (*Pan troglodytes*) and pygmy chimpanzee (*Pan paniscus*), differences in their respective neurological systems relating to empathy and aggression may underlie differences in their social behavior (Rilling et al. 2011).

Thus, while there is commonality in psychoneural processes in humans, primates, and other mammals, this need not translate into equivalent commonality for cognitive capacities and social behavior. In some areas, such as working memory, it appears that small, quantitative differences can translate into important qualitative differences. For example, cognitive processes such as recursive thinking and reasoning require a larger working memory than is generally found among the chimpanzees (Read 2008). Consequently, we will assume continuity at the bioneurological level while recognizing that differences of degree can lead to differences of kind. Our concern, then, is not so much with brain architecture and organization per se, but with possible changes in, or expression of, capabilities that may have arisen in the evolutionary pathway leading to modern *Homo sapiens*. Human behaviors are subject to biological constraints and evolutionary change through natural selection, but ascribing the expression of all human behaviors to evolution acting on a biological substrate requires assuming an unrealistically tight relationship between substrate and expression through behavior.

However, that is not all there is to the matter. The question at hand is not just one of the degree of shared traits between ourselves and other primate species, but whether the factors that have led to the social structure of non-human primate species are the same factors that have led to the social structure of human societies. This is the heart of the "quantitative but not qualitative" difference argument: do explanatory arguments that account for the properties of non-human primate species apply equally to human societies, so that the differences are primarily in the degree to which one factor or another is important in the human versus the non-human primate case?

Marriage as a Cultural Practice

Let us go back briefly to our matriline example to see that even this expanded criterion does not, by itself, provide homology between human societies and non-human primate societies. If the similarity is an analogy and not a homology as suggested above, then viewing matrilineality as providing a direct connection with the social structure of other primate species would not be valid. While the lineage structure based on mother-daughter links in a matrilineal society has similarities to an OW monkey society in which males leave the natal troop upon reaching sexual maturity, a striking difference arises with marriage—a universal aspect, in some form, of human societies. Marriage is a human practice in which the society, in effect, sets up rules about when an offspring produced through sexual intercourse will be accepted as a legitimate, social member of that society.[2] We can see this aspect of marriage through the terms we use, as culture bearers, for a child born to a male and a female who are not married to each other. These range from the more-or-less descriptive "born out of wedlock," to "illegitimate child," and finally to the pejorative "bastard," along with the negative connotations of bastard such as villain, rogue, snake in the grass, and good-for-nothing, to name but a few. The synonyms imply that the normal sense of one acting according to what is considered to be moral behavior is lacking in someone who is a bastard. It is as if the person born out of wedlock is an incomplete person, or at least an incomplete person in a social sense. One other synonym, the descriptive "whoreson," also brings into question the morality of the woman who gave birth to the child considered to be a bastard.

Marriage is a peculiar institution from the viewpoint of biological evolution since it has the potential to reduce reproductive fitness by making it more difficult to begin sexual activity leading to reproduction after reaching puberty, hence reducing fitness as measured by number of surviving offspring produced over an individual's lifespan. Unlike virtually any other, non-human mating system, it subjugates the interest of the individual to the interest of the group. Marriage restricts the individual from beginning to reproduce until he or she has first satisfied the concerns of the group in which he or she is a member regarding the status of the offspring that one may produce. A contract, even if implicit, is being made and enacted through a marriage ritual that involves not just the male and female to be married, but their kin, friends and others as representatives of the society of which they are a part. Marriage establishes the status of societal member for any offspring the couple may produce. It is as if each of the two persons being united through marriage is agreeing to, or has already agreed upon, the imposition of restrictions on his or her sexual and other behaviors in return for the right to have his or her offspring recognized as full-fledged societal members.

This can be seen in what the Tiwi, traditionally a hunter-gatherer group that lived on Melville and Bathurst Islands off the northern coast of Australia, say about sexual intercourse and pregnancy. They assert that while sexual intercourse leads to pregnancy and birth, sexual intercourse by itself does not produce a Tiwi child: "A Tiwi must be *dreamed* by its father, the man to whom its mother is married, before it can be conceived by its mother" (Goodale 1971: 138, emphasis in the original). Dreaming is a complex idea-system that is central to the process of constructing the social identity for a child-to-be and "is the catalyst that transforms a Tiwi from the world of the unborn to that of the living" (140). In overly simplified terms and using our vocabulary, the Tiwi make a distinction between producing a biological offspring who is a member of the species *Homo sapiens* and producing a Tiwi child who will have the social and kinship identity that is central to what it means to be a Tiwi.

Marriage as a Cultural Idea System

Marriage, then, is a cultural construct central to human societies and has no counterpart outside of human societies. One might attempt to consider marriage as just elaborated pair-bonding since marriage has functionalities that can be analyzed in the same way as a biological trait like pair-bonding. We can consider marriage as having the function of ensuring at least a temporary bond between a male and a female, thereby affecting their reproductive success. However, to reduce marriage to just this kind of functionality requires ignoring the way in which marriage is part of a cultural idea system. The cultural idea system has to do with what constitutes the culturally constructed social and kinship universe in which individuals are embedded and through which social identities are defined and taken on, along with the culturally recognized changes that may take place in one's social identity while going through a life cycle from birth to death. Kinship, as it is found in human societies, is a cultural construct, and a kinship terminology made up of the terms used to refer to relatives is not simply a linguistic device for expressing the biological relations we have, or we believe to have, with other members of our social group.

For some kinship terminologies, such as the terminology used by English speakers, there is a superficial resemblance between some of the kin terms and biological relations. For example, most English speakers consider the English kin terms mother and father to identify one's biological mother and biological father. To say, "She is my mother," implies that the speaker is asserting that the female in question is his or her biological mother. However, the same expression can also be used if the speaker was adopted. The kin term "mother" is used in cases of adoption despite the absence of biological connection. We could modify the term "mother" in cases of adoption as we do with remarriage where the already existing child of the father refers to the father's new wife as "stepmother," or the more recent use of the term "birth mother" to designate the biological mother of the child when she is not recognized as the mother in a kinship sense. The fact that we do not do so in the case of adoption indicates that the meaning of "mother" in our kinship terminology is not synonymous with "biological mother."

In other terminologies, even this superficial resemblance between kinship relations identified through a kinship terminology and biological relations disappears. For societies that have what is technically called a classificatory (or bifurcate merging) terminology, their term that is equivalent to our term "mother" applies equally to a biological mother, her sister, her female first cousins who are her biological mother's sister's daughters, her female second cousins who are her biological mother's biological mother's [grandmother's] sister's daughter's daughters, and so on. The sheer difficulty of biologically describing who is referred to as "mother" in cases like this underscores the extent to which the terminology in such societies is not built by analogy with biological relations.

This does not mean that kinship systems are cultural constructs that determine kin relations in a manner independent of biological relations, but only that kinship terminologies are not constructed according to the logic of biological relations (Read 2001). Consequently, the relations identified in a kinship terminology may, for reasons of logical consistency as part of a system of kin term concepts, take on a form that is even contrary to biological distinctions. For example, among the Crow Indians, who had a matrilineal society, their term used to refer to a biological father applies equally to that man's sister's son, that man's sister's daughter's son, and so on.

The assertion that kinship systems are grounded in the biology of reproduction only identifies the possible origin of some of the basic concepts of kinship that have to do with the positions making up the cultural concept of a family as a basic social unit, not the form that a culturally constructed system of kinship relations takes on. The order of culturally constructed kin relations is not the order of biological relations. Instead, as Lévi-Strauss (1969[1949]) pointed out, the appearance of human societies has to do with the formation of a new order, an order based on cultural rules and not just on biologically based behaviors or individual learning. This leads us to the second main idea about the transition to human societies.

Disconnect Between Ancestral Primate and Human Societies

Lévi-Strauss saw a disconnect between ancestral primate societies and human societies: "It seems as if the great apes, having broken away from a specific pattern of behaviour, were unable to reestablish a norm on any new plane. The clear and precise instinctive behaviour of most mammals is lost to them, but the difference is purely negative and the field that nature has abandoned remains unoccupied" (1969[1949]: 8).

The disconnect he saw with the great apes arises because the biological basis for behavior—what he calls a "specific pattern of behaviour"—gave way to increased individualization, leading to a breakdown in what had been social organization built around behaviors constrained through Darwinian evolution driven by natural selection. Lévi-Strauss built his argument on an observation made by the primatologist Robert Yerkes: "The orang-utan, gorilla and chimpanzee especially resemble man in this individualization of behavior" (Yerkes 1927: 181, as quoted in Lévi-Strauss 1969[1949]: 7). Consequently, the trajectory going from an ancestral primate species for our species would, in simplified form, be from coherent and integrated social systems dependent on regular patterns of behavior derived through biological selection, to loss of coherency and integration due to increased individualization among group members, and then back to coherent and integrated social systems, but based no longer on biologically grounded forms of social organization driven by natural selection. Thus, he argued, what arose during hominin evolution leading to our species, *Homo sapiens*, was a new order, one based on behaviors constrained by rules that identify norms and normative behavior expressed through roles, with the structure of role systems grounded in the conceptual systems we refer to as culture. For Lévi-Strauss, the presence of rule-constrained behavior is the sign of a culturally grounded social system: "Wherever there are rules we know for certain that the cultural stage has been reached" (8).

While some might be tempted to argue that the hand clasping of the Mahale chimpanzees while grooming is a kind of "rule," this would change the notion of a rule into a description of an already existing behavior, not the basis upon which behaviors are formulated. For Lévi-Strauss the rules are mental constructions that do not arise from, or merely reflect, already existing behavior. Lévi-Strauss argued that what is universal about incest concepts in human societies is rules restricting who can marry whom, while the content of what constitutes an incestuous marriage is highly variable and even inconsistent from one society to another. Although most societies prohibit a marriage between a brother and a sister, for example, some societies, such as traditional Tonga and ancient Egypt, institutionalized brother-sister marriage for the highest-ranking individuals in the society. Interestingly, Roman Egyptian society then accepted brother-sister marriage after Rome gained control over Egypt, as if the pharaonic brother-sister marriages legitimized such marriages for everyone. Coming closer to home, incest rules regarding cousin marriages vary by state in the United States, as discussed by Martin Ottenheimer (1996). Yet other kinds of rules apply to cousin marriages in other societies. In many societies, what anthropologists technically define as parallel cousins (cousins through same sex siblings in the parental generation, such as one's father's brother's child or one's mother's sister's child) are prohibited as marriage partners, yet cross-cousins (cousin through cross-sex siblings in the parental generation, such as one's father's sister's child or one's mother's brother's child) may be seen as preferred or even expected marriage partners. What is meant by "parallel cousin" and by "cross-cousin" is not, however, their biological definition but their definition through culturally defined kin relations. A cross-cousin, from this viewpoint, is a person referred to as child by someone either referred to as brother by a female referred to as mother or as sister by a male referred to as father by speaker. For groups such as the Tiwi where their term for mother (or for father) may refer to several females (or males) and the term for brother (or sister) may refer to more individuals than just one's biological siblings, and so on, a biological cross-cousin defined in this manner is only one kind of cross-cousin. From a kinship terminology perspective, the matter is much simpler. All those persons we technically define as cross-cousin from our outside perspective would just be referred to by a single kin term by members of the society in question. Just as our term "cousin" identifies for us a number of persons with different genealogical relations to us that we equally refer to by the term "cousin," they have a term that identifies for them those persons that anthropologists define technically as cross-cousins.

Continuity in Process, Discontinuity in Effect

Although Lévi-Strauss's argument serves to identify what happened during hominin evolution that makes human societies distinct from the kinds of societies and social systems we find in other primates, it leaves silent the crucial matter of how this would come about. Here we have an enigma. The evolution of our hominin ancestry away from other primate species must have been driven by biological evolution through natural selection. There is no special form of evolution involved in the changes that took place in the history of our ancestry. Our hominin ancestors did not, somehow, arise by some means other than the same biological evolution we invoke to account for the origin of, and change in, all other species. This would seem to imply that however different our species may appear to be (or not to be, depending on one's perspective!) when compared to other primate species, since our species came about through the same biological evolution used to account for the appearance of all other species, our species can only differ quantitatively, not qualitatively. Yet from what we know about human societies and the centrality of culture and cultural idea systems for understanding what it means to be human and how human societies are organized and operate, it is hard to refute Lévi-Strauss's argument that with the appearance of human societies our ancestors found a means to occupy "the field that nature has abandoned"; hence, what arose in hominin evolution was a new kind of society. In this new kind of society, change and evolution are no longer adequately expressed through just the logic of biological evolution driven by natural selection (Lane et al. 2009).

That such a change could and did occur, as well as a tentative explanation of how it occurred, is the essence of the argument presented in this book. The argument will be framed by considering two critical points separated in time. The first point is the kind and form of social structure and organization we find with many of the OW monkeys; that is to say, with forms of social structure and organization that we can properly account for by reference to the logic of biological evolution driven by natural selection. The second critical point is hunter-gatherer societies with their rule-bound behavior based on cultural concepts that permeate all aspects of the lives of the persons in these societies—as is true for us as well. Hunter-gatherer societies will be used for the second point for several reasons. Their small size—often the entire society may be at most five or six hundred persons—allows us to strip away the complexities of larger scale societies that relate primarily to their larger size. Their mode of subsistence is based on procurement of resources in a manner (in most cases) not designed or intended to increase the abundance of resources, which means we do not need to take into account the added complication of societal properties that arose once it became possible to exert control over the productivity, abundance and diversity of resources. (There are exceptions regarding not deliberately affecting the productivity of resources such as the use of range fires in Australia to increase undergrowth attractive to animals, but the specificity and form of the exceptions make these "exceptions that prove the rule.") The antiquity of the hunter-gatherer mode of resource procurement in which technology plays a central role goes back to the Upper Paleolithic and so to the transition from hominin societies where, at most, only the rudiments of culture are present to those societies in which the survival of the group became dependent on integrating the cultural dimension with all aspects of a group's life-ways.

From Ancestral Primate to Hunter-Gatherer
Forms of Social Organization

The question to be answered in this book is: how did we evolve from the first point to the second? To answer this question, we must begin with a biological perspective and initially only consider changes that fit within the biologically defined, evolution-by-natural selection framework. Then we must demonstrate how those changes could lead to the novel situation in which further change becomes increasingly divorced from the biologically constituted framework of change driven by natural selection. This requires that we identify, specifically, what was introduced, as part of the biological framework, that could then lead to the "new order," using Lévi-Strauss's words, that has been extensively described and analyzed through the detailed ethnographic study of human societies. Because our primate relatives had the same amount of evolutionary time from a last common ancestor as did our hominin ancestors, we must also be able to account for why such a radical change took place in our evolutionary ancestry and not in other primate species.

We have a time frame for this transition. Divergence from a last common ancestor with any of the living non-human primate species dates back to around seven to nine million years ago, and by the time of the middle Upper Paleolithic around 25,000 years ago, the foundations for what we understand as human societies—societies comparable in complexity and cultural embeddedness to even the smallest hunter-gatherer societies of today—were in place. We also know that another hominin—commonly known as the Neanderthals—survived as a distinct deme (some would say a different species, but that is not critical here) until around 35,000 years ago. We know that in what we now refer to as the Middle East, Europe and possibly eastward into central Asia, our ancestors and the Neanderthals overlapped in space and to some extent in time, especially after our ancestors expanded into the European area beginning around 40,000 years ago. What has become increasingly evident is that the demise of the Neanderthals was not due to inferior technology, limited hunting ability, or climatic conditions—that is to say, it was not due to extrinsic factors—but to differences in the cognitive abilities of our ancestors that enabled them to conceptualize and integrate themselves beyond the local residence group that was the center for day-to-day activities (Leaf and Read forthcoming). In effect, it was the ability of our ancestors to be able to begin to conceptualize like us; that is, to be able to construct conceptual systems that provided the means to integrate together what would otherwise be several disparate groups into a conceptual whole that gave them a competitive advantage concerning effective use of resources. This made it possible to expand the area over which resources could be procured by a single residence group; such groups were less affected by local variation in resource abundance, which enabled them to outcompete the Neanderthals for resources. In Lévi-Strauss's words, "the field that nature has abandoned" had now become occupied.

Hunter-Gatherers: The "Real People"

We know from modern hunter-gatherers that the conceptual means by which separate residence groups are integrated together, thereby enabling non-contentious movement of individuals (and families) from one residence group to another, is a culturally constructed system of kinship expressed concretely through the kinship terminology of a hunter-gatherer society. The members of a hunter-gatherer society see themselves as conceptually bounded and typically refer to themselves collectively as "We, the real people," or words to that effect. The !Kung san (a hunter-gatherer group in northwest Botswana) "refer to themselves as '*Zhun/ twasi,*' which may be sensibly glossed as the real people" (Konner 1972: 285). The Jahai of northern Malaysia refer to themselves as *menra* or "real people" (van der Sluys 2000: 433). The Beaver Indians of Canada called themselves *Dana-zaa* or "real people" (R. Ridington and J. Ridington 2006: ix). The Owens Valley Paiute named themselves *nümü*, meaning "the 'people'" (Steward 1933: 235). The word *Inupiat*, the name of an Inuit group in northwestern Alaska, means "the real people" (Gadsby 2004: 54). The Kusunda of Nepal refer to themselves as "*mihhaq* 'the people'" (Watters 2006: 9). These are just a few of the many examples of a hunter-gather group referring to its members as the "real people."

Day-to-day activities, the division of the whole population into residence groups, marriages, the sharing of resources, and the like take place in the context of the "real people," and interactions with outsiders are based on distrust and fear. According to one of Alfred Radcliffe-Brown's informants among the Kariera, a hunter-gatherer group on the western coast of Australia, "If I am a blackfellow [*sic*] and meet another blackfellow that other must be either my relative or my enemy" (Radcliffe-Brown 1913: 151). If we make an analogy with the OW monkeys, the relations between hunter-gatherer *societies* are like the relations between monkey *troops*. This indicates that the social domain of those interacting on a day-to-day basis has changed in scale by an order of magnitude as we go forward from the non-human primates to our ancestors.

Cultural Kinship and "Real People"

If there is categorization of people into social "real people" and dangerous strangers, then we need to know the basis for the categorization. Typically, it is based on the presence or absence of kinship relations. One's kin are the real people, and before social interaction can take place, one needs to know that the other person is one's kin. However, kinship is not biological kinship, but rather kinship as it is culturally constructed and based on a conceptual system that enables individuals to compute in a simple manner whether they are kin to one another. If A knows his or her kin relationship to B, and B knows his or her kin relationship to C, then A, B and C can each calculate their respective kin relationships to each other through the logic of their kinship terminology. As Steven Levinson (2002) comments regarding the group living on Rossel Island in New Guinea: "What is essential in order to apply a kin term to an individual X, is to know how someone else … refers to X. From that knowledge alone, a correct appellation can be deduced. For example, suppose someone I call a *tîdê* 'sister' calls X a *tp:ee* 'my child,' then I can call X a *chênê* 'my nephew,' without having the faintest idea of my genealogical connection to X" (18).

Change from Face-to-Face to Relational Systems of Interaction

If the boundary of those who are the real people and their internal social organization is based on a culturally constructed system of kin relations, then what was introduced in the evolution leading to our Upper Paleolithic ancestors and to their replacement of the Neanderthals must center on a change from the face-to-face interaction that characterizes non-human primate societies to a social system constituted according to a constructed system of kin relations. It is this transition that forms the core of the argument presented here for the evolution of human societies and, consequently, what makes our societies qualitatively different from non-human primate societies. The constructed system of kinship is not an elaboration upon, nor emergent from, a system of biological relations. Instead, it is the essence of what we mean by a cultural idea-system—a constructed (in the sense that it is not modeled on what is extrinsic to us) conceptual system with an internal logic that ensures common understanding and meaning for that system of concepts by those who share it in common through enculturation. In Lévi-Strauss's terms, it is a system of rules—rules about who are our kin and who are not, rules about how collectively we are organized together as a system of kin, and rules about how we interact with our kin. Once this system of rules is in place, evolution of human societies becomes evolution in the system of organization for a society, and evolution becomes constrained and directed by the logic and consequences of what has been constructed, not by change in individual traits per se (Read, Lane, and van der Leeuw 2009). It is no longer evolution driven just by changes in the transmittable traits of individuals and selection among variants in those traits, the hallmark of Darwinian evolution. Rather, it is evolution more in the manner envisaged by nineteenth century scholars— such as Lewis Henry Morgan, Edward Burnett Tylor, Emile Durkheim and Herbert Spencer—who saw human social evolution as change in the organization of human societies.

Nineteenth Century Evolutionists

Although the claim of a simple unilineal evolutionary sequence going from band to tribe to chiefdom to state associated with the nineteenth century evolutionists—but mistakenly linked to Spencer (Carneiro 1981)—has long been discredited, they correctly focused on evolution as directly relating to the organization of human social systems, rather than as change emerging from the evolution of individual traits. The weakness of their argument lies in their failure to relate evolution to the multiple modes of organization that characterize all societies (Leaf and Read forthcoming). Instead, they considered these broad societal forms as if each were a single institution or organization undergoing evolution. However, each societal form they identified is the summary classification made of the outcomes of multiple processes involving multiple institutions and organizations and does not address change in any specific mode of organization such as kin relations expressed through a kinship terminology. Accounts of change must refer to specific organizations and changes in their interconnections within a society, not to outcomes summarized as band, tribe, chiefdom or state levels of organization. There is no "law" that "causes" a band to become a tribe or a chiefdom or state. In addition, even if there were always a pattern of change from lower to higher levels of organization, as is assumed in the evolutionary sequence described earlier, we still would only have the rough equivalent of Kepler's *data model* of planets following elliptical orbits and not Newton's *theory model* of planetary motion derived from the relationship between force and motion (see Read 2008a for the role of data models and theory models in explanatory arguments).

Spencer recognized the difficulty rhetorically: "Is this law [of going from the relatively homogeneous and incoherent to the relatively heterogeneous and coherent] ultimate or derivative?... May we seek for some all-pervading principle which underlies this all-pervading process? Can the inductions [i.e., data models] ... be reduced to deductions [i.e., theory models]?" (1900: 368). He answered in the affirmative by arguing that the ultimate principle lies in the fact that even with homogeneity there can never be a permanent equilibrium as some differences will always arise: "The condition of homogeneity is a condition of unstable equilibrium ... an internal instability" (372–373). For hunter-gather groups, the instability can be seen in the fact that even with an ideology of equality and deliberate downplaying of differences in individual skills (Boehm 1999), some differences will always arise, and so leaders will emerge, thereby ending the homogeneity of equality. Because of this instability, he argued, evolution will take place as he indicated in his rhetorically expressed inductive law.

Although Spencer grappled with the problem of determining the first principles from which theory models could be deduced, his solution lacks the specificity needed to formally deduce theory models of human social evolution. More recent exponents of social and cultural evolution—such as Julian Steward, Elman Service, Marshall Sahlins, Leslie White, and Robert Carneiro—have related it to environmental conditions, which have been expanded by other evolutionists like Marvin Harris to include technology and, especially, material conditions. For all of them, social evolution has to do with change in the form of social organization, which already presupposes that a transition has taken place from non-human primate forms of emergent social organization to a new social order in which "social organizations are ... based on coherent and well organized systems of ideas ... that create and objectify patterned systems of behavior through cultural instantiation of idea systems" (Leaf and Read forthcoming: 17).

Notes

1. The hypothesis that our last common ancestor with a non-human primate species is also the ancestor of the chimpanzees, with social organization comparable to that of modern chimpanzees, has been examined in detail and argued to be more likely than any of the alternative hypotheses (Chapais 2008). Consequently, we will assume chimpanzees can be used as a model for the last common ancestor.

2. Marriage in some form is thought to be universal, although the Moso, a matrilineal ethnic group in the Yunnan and Sichuan Provinces in China, are said not to have any traditional form of marriage institution, and instead reproduction takes place in the context of *tisese* ('walking together') sexual relations between adults that can be initiated by either party (Shih 2010). Adulthood is marked by a *chaeji* ('wearing skirt') ceremony for girls and *hliji* ('wearing pants') ceremony for boys, that transforms them, twelve years after birth, from a state of being soulless into "real human beings" (2010: 250). On the eve of the day that the household ceremonies take place, all of the initiates from several villages "eat together and sing and dance through the night" (2010: 249). This ceremony with the subsequent *tisese* sexual relations is strikingly similar to the *tali* tying ceremony and subsequent *sambandham* sexual relations among the Nayar of India (see Gough 1959 for a description of the Nayar marriage system). While the Moso system is not a marriage institution as it is usually understood, the *chaeji* and *hliji* ceremonies have the function of legitimizing sexual relations leading to reproduction, hence any offspring that are produced are considered to be fully legitimate (2010: 76).

CHAPTER 2

The Primate Beginning Point

For our odyssey we need a beginning point that is unambiguously on the biological side of the biological and cultural divide between non-human primates and ourselves. We also want the species making up the beginning point to be already part of the phylogenetic trend towards more complex cognitive capabilities and new forms of social organization and social structure that arose during the evolutionary trajectory leading to modern *Homo sapiens*. The social systems for these species should be complex enough to contain evolutionary roots for the advanced cognitive abilities and complex forms of social organization that characterize our species. At the same time, the social systems should be in a position, phylogenetically speaking, where the relationship between the major features of social organization and social behaviors can still be explained by reference primarily to biological evolution driven by natural selection in its various forms (individual selection, sexual selection, and biological kin selection). Our beginning point should thus provide us with a baseline showing the level of complexity—and limitation—of social organization explicable by reference to natural selection as the primary driving force for evolutionary change within a species.

These constraints lead us to the OW monkeys. In keeping with our requirement for a beginning point, the OW monkeys are part of a trend toward the cognitive and social complexity that characterizes our species as we have evolved to have cognitive abilities not shared with other mammals. The New World monkeys will be excluded since common ancestry of *Homo sapiens* with extant OW monkey species traces back to the Oligocene (about 35 million years ago) and so postdates the evolutionary divergence between the Old World and New World monkeys. Thus, the New World monkeys are separated from the evolutionary trajectory leading to *Homo sapiens*.

The basis for the cognitive distinctive of the OW monkeys in comparison to other mammals can be seen (Figure 2.1) through prolongation of an internally clocked time span for body growth from zygote to adult form in the anthropoid primates (Vinicius 2005). Virtually all animal forms of life ranging from crustaceans to birds and mammals have the same growth pattern (West, Brown, and Enquist 2001), except the anthropoid primates. The anthropoid primates have, uniquely, a prolonged time span for growth. The prolonged growth time span in the OW monkeys, and even more so in the great apes and our species (see Figure 2.1), implies that there is more time for brain development and maturation of brain processes in comparison to other mammals. This fosters the formation of behaviors based on more complex cognitive processes and has enabled, in a phylogenetically systematic manner, increasingly complex behaviors to become part of the behavioral repertoire found within primate species going from OW monkeys to ourselves.

As shown by the sequence of curves in Figure 2.1, the prolongation of brain and body development is most extensive in *Homo sapiens* within the anthropoid primates. This prolongation is also coupled with a faster brain growth rate for our species in comparison to other primates (Robson and Wood 2008), which leads to the high degree of encephalization (ratio between brain size and body size) in our species. Preceding *Homo sapiens* in this sequence are the great apes and, before them, the OW monkeys. This phylogenetic trend implies that we can consider the OW monkeys to be within the initial part of a general trend among the anthropoid primates towards forms of social organization and social structure that increasingly reflect capacities derived from the evolution of more complex cognitive processes. Their cognitive capacity for more complex forms of social organization than is the case for most other mammals has enabled the OW monkeys to adapt to virtually every set of climatic and ecological conditions on Planet Earth. By selecting the OW monkeys to be the beginning point, we begin our odyssey with adaptations to the same range of climatic and ecological conditions as the hunter-gatherers that make up the ending point for our odyssey but, as we will see, with very different forms of social organization.

Baseline Pattern for Social Organization

We begin with general characteristics of the social organization of the OW monkey. Broadly speaking, the daily life of the members of an OW monkey species is based on a group, generally referred to as a troop,

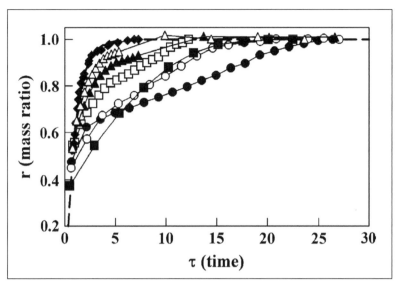

Figure 2.1 Comparison of primate growth curves to a general growth pattern (dashed line) that characterizes animal forms of life ranging from crustaceans to mammals. The vertical axis is given by $r = (m/M)^{1/4}$, where m is body mass at age t from birth and M is adult body mass. The horizontal axis is given by $\tau = at/4M^{1/4} - ln(1 - (m_0/M)^{1/4}$, where m_0 is birth weight and a is a cell metabolism parameter. (See West, Brown, and Enquist 2001 for details.) The curves from left to right can be divided into four groups showing the general growth pattern for: (1) cows (◆), (2) monkeys: Goeldi's monkey (△), pygmy marmoset (▲), rhesus monkey (□), (3) great apes: gorilla (■), chimpanzee (○), and (4) human (●). *Reprinted from Vinicius 2005: Figure 2 by permission of Elsevier.*

consisting of adult males and females plus juveniles and infants, with the sex ratio for adults generally skewed towards females. The primary features of a troop's social organization have been explicated by reference to ecological factors, as these are "sufficient to explain variation on group size, female dispersal, and establishment of hierarchical dominance relationships" (Izar 2004: 95). These factors can be incorporated into a general, ecological model for relations among females in primate species (see Figure 2.2). The model was developed initially by Richard Wrangham (1980) to account for female- versus male-bonded species, was later modified by Carel van Schaik (1989) to take into account a qualitative distinction in female-bonded primate species for egalitarian versus nepotistic forms of dominance hierarchies, and was further modified by Elisabeth Sterck and colleagues (1997) to account for the social effect of infanticide (when it occurs) on female social organization.

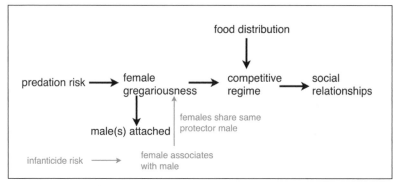

Figure 2.2 Causal ecological and social model for the formation of female social relationships in primate species. Predation risk leads to formation of groups coupled with female gregariousness. Relative degree of competition for access to food resources within and between groups of females leads to different patterns of female social relationships. When present, male infanticide adds another dimension (shown in gray) to the factors affecting female social relationships. *Diagram based on Sterck, Watts, and van Schaik 1997: Figure 6.*

The model includes two primary causal factors: (1) predation risk leading to group formation and gregariousness among females and (2) the pattern of food distribution affecting the extent to which there is female competition over food resources. The consequences of these causal factors are expressed ultimately through female social relationships. Social relationships in the model are characterized by contrasts between egalitarian or nepotistic behaviors, female resident or female dispersal, and tolerant or despotic hierarchies, thus leading to eight possible forms of social organization of which only four are realized empirically (see Figure 2.3). Secondary aspects of social relationships such as the presence of coalition formation by females in agonistic encounters and the presence of linear dominance hierarchies should co-vary, according to the model, in a regular manner according to the location of a species on these three dimensions. The authors show that the actual patterns for female social organization in primate species, ranging from prosimians to the great apes, empirically fit their model quite well (see the distribution of species in Figure 2.3).

Consider the model in more detail. According to the model, group formation in response to predation risk leads to competition over access to resources. Competition may be between groups and/or within a group. For example, in some species, such as the vervet monkeys, a troop may be highly territorial (Cheney 1987) and defend its feeding

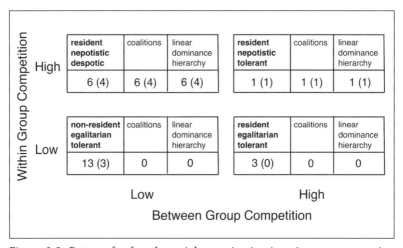

		resident nepotistic despotic	coalitions	linear dominance hierarchy		resident nepotistic tolerant	coalitions	linear dominance hierarchy
High		6 (4)	6 (4)	6 (4)		1 (1)	1 (1)	1 (1)
		non-resident egalitarian tolerant	coalitions	linear dominance hierarchy		resident egalitarian tolerant	coalitions	linear dominance hierarchy
Low		13 (3)	0	0		3 (0)	0	0

Within Group Competition

Low   High

Between Group Competition

Figure 2.3 Pattern for female social organization in primate taxa ranging from prosimians to great apes. Data are the number of taxa that fit into each category. All taxa are female philopatric except those categorized as "non-resident." Numbers in parentheses are for OW monkeys only. Two OW monkeys (*Colobus guerza* and *Presbytis entellus*) were classified by Sterck, Watts and van Schaik 1997) as Resident/Egalitarian (?) and are not included (neither has coalitions nor a linear dominance hierarchy). Variation in the intensity of competition for food resources, expressed both within and between groups, accounts for the primary patterns of female social relationships. *Data are from Sterck, Watts and van Schaik 1997: Table 2.*

area against other troops of the same species. Other species, such as the baboons, are not territorial, and there may be overlap of feeding ranges, but with the overlap separated in time.

In either case, encounters between troops are generally marked by antagonism or avoidance, thus making the troop a more-or-less closed social unit, except in one key aspect. In virtually all OW monkey species, one sex or the other (depending on the species) transfers from natal troop to another troop at time of sexual maturity. Among the cercopithecine (seed, tuber, shoot eating) species, it is the adolescent males who leave the natal troop. The colobine (leaf-eating) species have a more mixed pattern wherein sometimes it is the females of a species that make the transfer. (The term *philopatric* is used for denoting forms of social organization in which one sex stays essentially for life in its natal troop.)

The prevalence of female philopatry has been explained through positing that when a key resource can be defended (such as a patchily distributed resource), then social organization should revolve around a stable core of matrilineally related females, and the females should be

the philopatric sex (Wrangham 1980). Philopatry has the genetic consequence that the individuals of the sex remaining in the natal troop will
be made up of biological mother/biological daughter lines in the case
of female philopatry and of biological father/biological son lines when
there is male philopatry. In either case, philopatry increases the genetic
relatedness among the individuals staying in the natal group. For the
sex that is transferring to a different troop, genetic relations among the
adult members in the receiving troop can vary from little or no patterning (when genetically related individuals transfer independently to new
troops) to patterns similar to genetic lines (when biologically related
individuals and/or members of a single natal troop repeatedly transfer
to the same receiving troop, as may occur in some species with male
transfer [Bernstein 1991, and references therein]). Philopatry by a single
sex has important genetic consequences, ranging from reduction of
inbreeding (though some mammals reduce inbreeding without philopatric forms of social organization [Moore 1992]) to forming conditions
under which biological kin selection becomes increasingly important:[1]
"some of the strongest evidence of [biological] kin selection in action
has been found among female Old World cercopithecine monkeys, such
as baboons, macaques, and vervet monkeys" (Strier 2000: 127).

Biological Kin Selection and Female Social Organization

There can be selection for social behaviors directed towards biologically
related individuals even absent any cognitive awareness of who are one's
biological kin as long as current behaviors positively bias the formation
of interacting dyads towards genetically related pairs of individuals; i.e.,
towards biological kin. Examples of such biasing behaviors include biological mother/offspring nursing or being raised as litter-mates, but the
latter tends to have limited effect for the anthropoid primates due to
inter-birth spacing and time to sexual maturity that generally precludes
infants other than twins being raised together. However, even absent
formation of dyads of biologically related individuals on a regular basis
through behavior patterns, there still can be selection for social behaviors directed towards biological kin. Philopatric females reside with
other, genetically related females, and so any interaction among female
group members is biased towards more frequent interaction among
genetically related females than would occur without female philopatry. Conditions like these persisting over several generations favor the
spread of biologically based, social behaviors by troop members through
biological kin selection even if a behavior does not have an immediate,

direct reproductive fitness benefit to the acting individual, such as grooming another group member. (While male philopatry has the same structure, the effects of biological kin selection may be less pronounced since male philopatry need not lead to stable patterns of interaction among biologically related males due to paternity uncertainty.)

By biological kin selection is meant selection for altruistic behaviors; that is, behaviors directed towards biological kin despite a reproductive fitness cost for the acting individual. Biological kin selection depends on the social behavior entailing sufficient, compensating reproductive fitness in the target individual, taking into account the degree of genetic relatedness between the acting and the target individual.[2] As noted above, biological kin selection need not occur just between already formed dyads of interacting, genetically related individuals. It can also arise under conditions such as philopatry where social behaviors are likely to be directed towards a biological kin since philopatry leads to genetically related individuals living together.

Engaging in social behaviors can reduce reproductive fitness; for example, opportunity costs lost due to the time and energy spent directing a behavior towards another when no direct individual fitness benefit is received in return. Many social behaviors are of this kind. A number of studies have shown that under female philopatry, females engage in social behaviors without direct reproductive benefit for the acting female such as grooming, time spent near other females, coalition formation, and tolerance of other females when feeding, all according to the acting female's degree of genetic relatedness to the receiving individual (see references in Silk 2002). The time and energy spent in such behavior is an opportunity cost since that time and effort could have been spent in behavior that has a direct and positive reproductive fitness benefit.[3]

Although usually expressed through the behavior of one individual directed towards another biologically related individual, biological kin selection can be viewed equally from the perspective of an ancestral individual and the size of the cohort, across generations, of individuals who trace back genetically to that ancestral individual. For anthropoid primates, life span is long enough to make it likely that an ancestral female will have living offspring and "grand"-offspring. Her net reproductive fitness benefit derived from social behaviors she transmits genetically is measured by the size of her cohort of offspring and grand-offspring in comparison to what would have occurred without the transmittal and/or expression of those behaviors. The specifics of how and which offspring interact with what other offspring according to their degree of biological relatedness can be considered summarily by correlating the size of the surviving cohort of progeny the ancestress engenders with the social behaviors she and her progeny engage in. Long-term studies on savannah baboons, for example, have demonstrated a positive relationship between size of the surviving cohort (measured as the number of offspring surviving at least one year) and the degree to which adult females (hence biological mothers) engage in social behaviors (measured by proximity to, being groomed by, and grooming of other adult females) (Silk, Alberts, and Altmann 2003). All together, biological kin selection appears to play an important role in the presence and form of social behaviors in the OW monkeys.

Emergent Forms of Social Organization

The conditions under which we can expect selection for social behaviors through biological kin selection are further enhanced in female philopatric OW monkey species through a social structure built around a long-lasting, linearly organized, transitive dominance hierarchy. The dominance hierarchy can be stable for long periods of time: "Although many of the individuals [in the baboon troop] have changed, the same two matrilines have occupied the top two positions for decades" (Cheney and Seyfarth 2007: 69). Changes in the dominance hierarchy relate mainly to demographic events such as births and deaths (Fairbanks 2000). As a result, a troop with a dominance hierarchy has a stable social structure that gives it long-term cohesion, even taking into account short-term upheavals such as an arriving male engaging in infanticide in some species when a male transfers in from his natal troop.[4]

From a dynamic viewpoint, a female dominance hierarchy incorporates a newborn female offspring as she matures, first as an infant extensively groomed by her biological mother and biological aunts and later through the support she receives from her biological mother when the infant challenges other adolescents with dominance rank below her biological mother, including her older biological sisters. Then, as she gets older, she learns to challenge on her own all other adolescents with dominance rank below her biological mother. This process leads to her being "slotted" into the dominance hierarchy immediately below her biological mother and above her older biological sisters, as well as above all other females with dominance rank below her mother (see Figure 2.4). Note that no intentionality by the biological mother is required for this process to generate the dominance hierarchy. She need only follow a simple rule: provide extensive social support including intervention in agonistic interactions involving your most vulnerable female offspring.[5] This rule can be activated simply by the tendency, for example, of female Japanese macaques to protect their younger offspring (Kawamura 1965). With this rule, the newborn daughter will learn that she is dominant to all females with ranking below her mother, and all females subordinate to her mother will learn that the maturing daughter is dominate to them because of the support she receives from her biological mother.[6]

The consequence of the rule is the emergence of a matrilineal social unit in which dominance ranking is based on biological relatedness across generations and in reverse age order within the same generation for a matriline. By being ranked above her older biological sisters, a female is also "shielded" from dominance challenges by females outside of the matrilineal unit. Consequently, the dominance ranking is not determined by the physical size and/or strength of females, but by biological kin-based social relations within a matrilineal unit and the dominance relation between matrilineal units. A female can know her dominance ranking vis-a-vis all other females in the group simply by

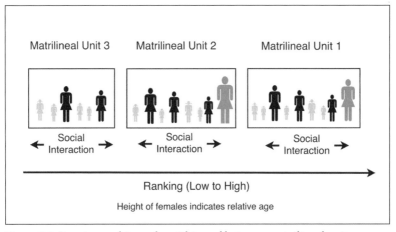

Figure 2.4 Dominance hierarchy within and between matrilineal units emerges from and is continued across generations by a new-born daughter eventually occupying a dominance rank just below her biological mother (see the reverse age-ranked positions of the gray females to the left of their biological mother in black). The same pattern holds across generations (see the daughters in black of the oldest females in dark gray and their ranking). Matrilineal units will be ranked vis-a-vis each other in accordance with the age of biological sisters in different units (see females in dark gray in Units 1 and 2) and ranking continues after the death of the oldest female (see Matrilineal Unit 3 without any dark gray female).

knowing her ranking within her matrilineal unit and the ranking of her matrilineal unit in comparison to other matrilineal units.

The matrilineal social unit formation model also implies that this form of social structure and dominance hierarchy is invariant with group size, since group size can be accommodated simply by adding more matrilineal units to the dominance hierarchy. The complexity, C, of a female primate's social world regarding dominance related behaviors is determined roughly by $C = f_{MU} + n_{MU}$, where f_{MU} is the number of females in her matrilineal unit and n_{MU} is the number of matrilineal units. When the species as a whole has a stabilized population size, an increase in group size primarily occurs through increasing the number, n_{MU}, of matrilineal units, since the magnitude of f_{MU} is determined by demographic factors, which are constant with a stabilized population size for the species as a whole. Keeping f_{MU} fixed and doubling the number, n_{MU}, of matrilineal units has only a relatively small effect on the cognitive complexity, C, for dominance-related behaviors in comparison to the magnitude of the increase in the total number of females in the group. Without the hierarchy of matrilineal units, the social complexity would scale with the number of females in the group. Eventually, though, an increase in group size can make the group as a whole less cohesive (Lehmann, Korstjens, and Dunbar 2007a), and group fissioning may then occur (Sterck, Watts, and van Schaik 1997).

Empirical support for modeling social structure as invariant with troop size is provided by a study of female grooming in two troops of olive baboons (*Papio cynacephalus anubis*) with 80 and 40 members, respectively (Sambrook, Whiten, and Strum 1995). Despite the difference in size of the two troops, the researchers found that there was neither a difference in the size of the grooming network nor in the grooming frequency by individuals in each of the two troops, leading them to conclude that the baboons were "'living in a small troop' of their own construction" (Sambrook, Whiten, and Strum 1995: 1679). The authors go on to observe that if individual baboons were interacting with all individuals in a troop size of 80-plus baboons, the troop would likely be behaviorally and cognitively too complex to maintain itself.

We also find that the matrilineal social unit model is consistent with the pattern for the presence or absence of linear dominance hierarchies in the Sterck and colleagues (1997) model for female social relations. Under the condition of high within group competition, rates of agonistic behaviors are also high, thus leading to higher rates of coalition formation that serve to resolve agonistic encounters. This leads to higher rates for the intervention by biological mothers in agonistic encounters involving their daughters, hence reinforcing the position of daughters in a dominance hierarchy. Taken together, these effects lead to stable linear dominance hierarchies. Conversely, with low within group competition coupled with resident females, the rates of agonistic encounters are infrequent and so there would be few opportunities either for the learning aspect of a dominance hierarchy by biological daughters or for conditions under which the stabilizing effect of a dominance hierarchy would be involved. Finally, the conditions that lead to the absence of female residence do not lead to establishing a dominance hierarchy amongst females.

When conditions lead to a stable dominance hierarchy, a female's rank within a matrilineal unit begins after birth when other females in her matrilineal unit (hence females who are biologically closely related to her) attempt to hold her, and continues with extensive grooming directed towards her by her biological mother and her biological maternal aunts (see Figure 2.4). Grooming is thus multifunctional, ranging from the pragmatic removal of ectoparasites to the working out of social relations among the individuals being groomed. A widely used model for the grooming pattern among female monkeys developed by Robert Seyfarth (1977) and supported by a number of studies (Schino 2001 and references therein), posits that the distribution pattern of female grooming in female philopatric species is explained by "individuals attempting to groom (1) members of their maternal lineage, (2) those of high rank, and (3) those with young infants" (Seyfarth 1980: 799). Because of competition among females for grooming partners, most grooming should occur between females of adjacent ranks (Seyfarth 1980), although females of low rank would also attempt to groom females of higher rank. Thus, high ranking matrilineal units will tend to have inwardly directed grooming, and grooming by members of low ranking matrilineal units will be both inwardly directed and directed outwardly towards higher ranking females (see Figure 2.5).[7]

Grooming, Social Organization and Neocortex Ratio

As discussed above, grooming is central to the formation of female dominance hierarchies through creating bonds among kin related females, starting from the birth of a daughter and continuing throughout the life span of females. Not surprisingly, females with new offspring are the object of grooming (Boccia 1998), and much of the grooming by adult females is directed to their offspring and other juvenile and infant biological kin (S. Gouzoules and H. Gouzoules 1987). More broadly,

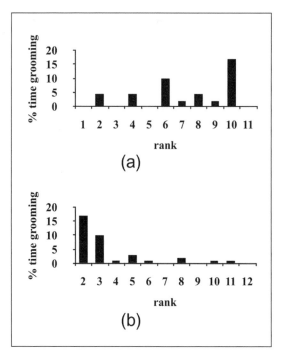

Figure 2.5: Percent time grooming by chacma baboons (*Papio cynocephalus ursinus*): (a) grooming by rank 12 female and (b) grooming by rank 1 female. Histograms illustrate the more varied grooming pattern of a low ranking female in comparison to a high ranking female. *Redrawn from Barrett and Henzi 2002: Figure 1, by permission of Brill.*

grooming is an important social mechanism for reinforcing group coherence in the face of potentially divisive factors such as competition over food resources and sexual competition between and among males and females (Dunbar 1988). Grooming plays, then, an important social function by facilitating and reinforcing interactions among adult females, with implications for coalition formation, resolution of agonistic encounters, and the like. The positive social functionality of grooming requires, however, a tradeoff against time that could be spent in other activities (through what some have referred to as market forces [e.g., Barrett et al. 1999]), hence there is also downward pressure limiting the time spent grooming. Consequently, species with higher rates of grooming should reflect conditions in which investment in the positive functionality of grooming has been favored.

We can investigate this hypothesis empirically by examining the frequency with which grooming rates occur over a wide variety of OW monkey species. When we do this, we find that the frequency distribution for the percentage of time spent grooming divides into two distinct modes (see Figure 2.6A). We can refer to these as low and high grooming rate modes. The numerical boundaries are: low rate < 6% and high rate

Chapter 2

> 7.5%. Next we determine if the modes correspond consistently with behavioral differences among these species. Figure 2.6B shows that there is an almost perfect association between grooming rate mode and the characterization of a primate species as territorial or arboreal. The single exception to the pattern is the Tana crested mangabey (*Cercocebus galeritus*): with a 5.5 percent grooming rate they are yet considered to be terrestrial. However, because they feed "at all heights in the canopy," their range is "restricted to gallery forest" (Homewood 1978: 380, 388), and their "semi-terrestrial behaviour enables the mangabeys to exploit arboreal food sources" (Homewood 1975: 58), it is suggested that theirs is more of an arboreal than a terrestrial adaptation and that they should be grouped under arboreal. If so, then there are no exceptions.

We also might expect, at first glance, to find a positive correlation between group size and grooming rates, since larger groups have

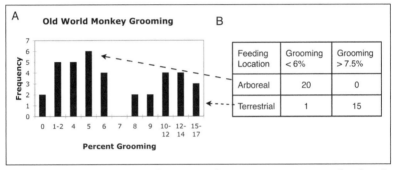

Figure 2.6 (A) Histogram of percent of time spent grooming for female philopatric species. Distribution is bimodal. (B) Modes correspond to arboreal primates with low grooming rates and terrestrial primates with high grooming rates. Species, group size, philopatry, location and % grooming are from Lehmann, Korstjens, and Dunbar 2007a: Table 2, except as follows: *Cercopithecus aethiops* (group size, Barton 1996; % grooming Seyfarth 1980); *Cercopithecus mitis* (% grooming, Cords 2002); *Erythrocebus patas* (group size, Barton 1996; % grooming, Chism 2002); *Macaca fascicularis* (group size, Umapathy, Singh, and Mohnot 2002); *Macaca radiata* (group size, Barton 1996; % grooming, Dunbar 1991); *Macaca sylvanus* (group size, Barton 1996; % grooming, Menard 1997); *Papio anubis* (group size and % grooming, Dunbar 1992); *Papio cynocephalus* (group size, Barton 1996; % grooming, Hill et al. 2003); *Presbytis entellus* (philopatry, Koenig and Borries 2001; location, Yoshiba 1967); *Presbytis johnii* (group size and % grooming, Poirier 1969); *Presbytis pileatus* (group size and % grooming, Stanford 1991); *Presbytis rubicunda* (location, Suprianata, Manullang, and Soekara 1986); *Semnopithecus entellus* (group size and % grooming, Sayers 2008); and *Trachypityhecus francoisi* (all data, Zhou et al. 2007).

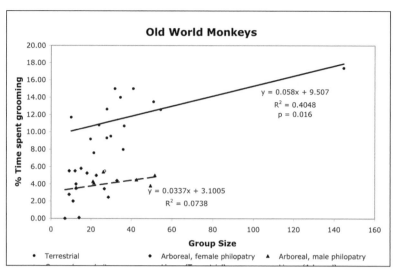

Figure 2.7 Plot of percent of time spent grooming versus group size. The correlation between group size and percent of time grooming for the terrestrial species is statistically significant only because of the outlier, *Theropithecus gelada*, on the right side of the graph. The weak correlation between the two variables for the arboreal species is not statistically significant.

more females and thus potentially a more complex social environment requiring more grooming to maintain group cohesion. Further, grooming rates among females should differ between female versus male philopatry since the two patterns have substantially different implications for the biological relationships among adult females in a group. However, these expectations are only weakly supported at best. When we compare percentage of time spent grooming to group size, controlling for terrestrial versus arboreal (see Figure 2.7), the moderate correlation for the terrestrial species is due solely to the extreme right data point (*T. gelada*), and without this data point the correlation between group size and percentage of time spent grooming is not significant statistically. The regression line for the arboreal species is also not significant statistically, and whether the species is female or male philopatric does not appear to make a difference in the pattern for the arboreal species.

For female philopatric species (which include most terrestrial species), the absence of a correlation between group size and rate of grooming can be accounted for by the size scale's independence of a nepotistic dominance hierarchy (see Figure 2.4) discussed above.

Female grooming rates should not vary with group size since the bulk of female grooming is within her matrilineal unit, and the size of a matrilineal unit is a consequence of internal, demographic factors and, thus, independent of the total number of females in the group outside of the matrilineal unit. Consequently, social complexity does not increase, or at most only increases slightly, with group size when there is a nepotistic dominance hierarchy. For the male philopatric species, the conditions favoring male philopatry—low between and within group competition among females over resources—lead to egalitarian relations among females (see Figure 2.3), hence the social group does not become more complex for females as the group size increases.

The terrestrial versus arboreal difference in grooming rates seems to reflect both a higher risk of predation due to feeding in the open and greater competition for food resources for the terrestrial species. Together, these conditions would favor mechanisms for maintaining social cohesion through grooming and so the higher rates of grooming among the terrestrial species. We can relate this pattern to the "social brain" hypothesis (Dunbar 1998), which posits a positive relationship between the neocortex ratio (the ratio of the cortex volume to the brain stem volume, hence an index of the degree to which selection has favored proportionately greater neocortex volume in the brain) and social complexity. According to the social brain hypothesis, the neocortex ratio for terrestrial species should be greater than for arboreal species. Actually, this is the case. The mean neocortex ratio for the arboreal species, $\bar{x}_{arboreal} = 2.29$, is smaller than the mean ratio for the terrestrial species, $\bar{x}_{terrestrial} = 2.57$. The difference is statistically highly significant, even with the small sample sizes in this data set.[8] In addition, the neocortex ratio does not correlate with group size in either the arboreal or the terrestrial monkeys, as is expected given that social complexity within these two groupings does not correlate with group size (see Figure 2.8).

The terrestrial species, then, are dealing with a more complex set of behaviors where the added complexity does not arise from group size but from external conditions that lead to more intensive rates of interaction among matrilineally related females. Consequently, we can view the contrast between the arboreal and terrestrial species as indicating that intensification of grooming is the primary means available to the OW monkeys for dealing with more complex social contexts. However, as pointed out by Dunbar (1992), intensification of grooming is ultimately limited by the competing demands made on the limited daily

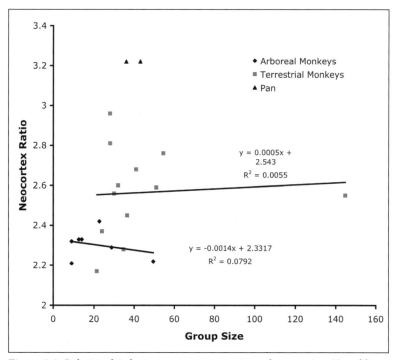

Figure 2.8 Relationship between neocortex ratio and group size. Trend lines for arboreal (♦) and terrestrial (■) monkeys are not significant. The two data points for Pan (▲) are for *Pan paniscus* (left data point) and *Pan troglodytes* (right data point). The trend lines are not statistically significant even with the presence of an outlier in the terrestrial OW monkey data. *Data are from Kudo and Dunbar 2001: Table 1.*

time budget available to females for the various activities and behaviors they engage in as part of their adaptation.

Summary

The OW monkeys provide a good starting point for our odyssey as they have developed adaptations through natural selection that incorporate greater social complexity than is found in most other mammals, while maintaining group coherency. A group is composed of males, females and offspring and is a social unit within which individuals go about their daily behaviors. As indicated by the ecological-constraints model (Milton 1984; Janson 1988; Wrangham, Gittleman, and Chapman 1993; C. Chapman, Wrangham, and I. Chapman 1995; C. Chapman and I. Chapman 2000), the optimal size of the group is in accord with ecological conditions measured by day range and food availability, though for demographic or other reasons such as predation risk, the group size may expand occasionally beyond an optimal size. When this happens and the group size reaches a point where social cohesion begins to break down, the group may split into smaller units along matrilines (Henzi, Lycett, and Piper 1997). Within this framework, biological kin selection appears to play an important role in selecting for social behaviors: fitness benefits accrue from social interaction among a female and her progeny and not just from increased reproduction; that is, with the OW monkeys a group is not only a means to cope with external, ecological conditions, but also a context within which there are fitness benefits and losses worked out through the pattern of social behaviors and social relations that have evolved. In this sense, they are more similar to our species than to other mammalian species. At the same time, we can also see limitations on the range of social behaviors that can be introduced under biological selection alone. One such limitation occurs with reciprocal altruism.

By definition, reciprocal altruism involves repeated, altruistic interactions between a pair of individuals where each party to the interaction benefits from the altruistic behavior of the other. Individually altruistic social behaviors are already present in OW monkey groups, and so we might expect instances of reciprocal altruism to have evolved. For example, individual A altruistically grooms a more dominant individual B, who benefits from the grooming through the removal of ectoparasites. Similarly, a more dominant individual E forms a coalition in an agonistic encounter between individuals C and D with the more dominant of C and D, who then benefits from that coalition. Nonetheless, putting the behaviors together so that A grooms E and reciprocally E forms a coalition with A seems to be beyond the capacities of the OW monkeys, as there is only a very weak correlation, r = 0.15, between grooming and coalition support in agonistic encounters (Schino 2006). From this we may conclude that there are limitations on social behaviors that can be introduced through biological kin selection, at least for the OW monkeys.

For the OW monkeys, the effectiveness of biological kin selection depends on non-cognitive mechanisms biasing repeated interaction towards biologically related individuals. When this occurs, biological kin will be the primary beneficiaries of social behaviors. Consequently, a pattern of extensive social interactions should not be introduced through biological kin selection when interacting with group members outside of one's matriline unit. Instead, and in accordance with the dominance relations between matriline units, behavioral encounters between females in different units generally reduce to dominance-submission behaviors. When this does not happen and there is an agonistic encounter, there may be a third-party intervention by a female dominant to the interacting pair of females. This female forms a coalition with the more dominant of the two interacting females and thereby resolves the agonistic encounter.

Dominance-submission behaviors are just part of the range of interactions that occur among the members of a matriline unit. Interactions may involve less biologically constrained, novel behaviors that, by being novel, have the potential for social disruption. Social cohesion is maintained, though, by keeping novel social behaviors within a matriline unit. The cohort of adult females making up a matriline unit is small, hence amenable to intensive face-to-face interaction through which social relations incorporating novel behaviors can be worked out.

All together, the restricted behavioral repertoire between members of different matrilines and the behaviors among a small cohort of adult females making up a matrilineal unit keep small, in comparison to the size of the group, the total repertoire of behaviors with which individual primates learn to cope as part of their social interactions. This small repertoire keeps manageable the cognitive task of forming expectations about the behavior of other group members, a key aspect of what makes a group work effectively as a socially coherent unit.

The small size of the cohort of adult females making up a biological matriline makes it possible, from a time-budget perspective, to engage as needed in intensive interaction through grooming and other ways for working out potential conflicts that might arise through "misunderstood" behaviors or through individual interests that may be in conflict. In these ways behaviors are called upon to deal with short-term contingencies, not long term goals.

Notes

1 Inbreeding reduction requires that less breeding take place between biologically closely related individuals than would occur under random mating, but does not depend on individuals being cognitively aware of who are
their biological relatives if there are behaviors that make it more likely for
genetically related individuals to repeatedly interact. For mammals, an offspring "knows" who is his or her biological mother simply through the facts
of birth and nursing, whether or not the offspring cognizes her as "mother"
in contrast to other females. Consequently, selection can favor biologically
based biological mother and biological son inbreeding avoidance by simple
"genetic rules" such as "do not breed with the female who has nursed you." In
other words, inbreeding avoidance can be implemented by taking advantage
of already present behaviors that favor dyads of interacting and biologically
related males without first requiring any ability to cognize another individual
as a biological relative and then acting according to that cognitive awareness.

2 According to Hamilton's rule, biological kin selection favoring a genetically
based, social behavior can occur when $c < rb$, where c is the reproductive
cost to the actor of doing the behavior, r is the degree of genetic relatedness
between actor and target, and b is the reproductive benefit gained by the
target from the actor's behavior.

3 More precisely, according to Hamilton's rule the gene in question will
increase in frequency when its expression underlies behavior directed
towards a biologically related individual for whom $c < rb$, where r is the
degree of biological relatedness, c is the fitness cost of doing the behavior
and b is the fitness benefit for the recipient of the behavior. For example,
the genetic relatedness between siblings is $r = 1/2$, and so Hamilton's rule
indicates that even a gene whose expression leads to an extreme altruistic
behavior, such as forfeiting one's life ($c = 1$), will increase in frequency in a
population when the altruistic behavior is directed towards one's siblings
and results in the survival of two or more siblings ($b \geq 2$) who otherwise
would not have survived.

4 Infanticide accompanying male transfer has been well-documented among
the langurs and appears to be an example of sexual selection for which male
fitness is increased through shortening the amount of time before females
re-start ovulation (Borries et al. 1999). His infanticide act has the consequence that he will produce offspring more quickly and with more certainty
(since he is also subject to displacement by other males) than would otherwise be the case.

5 For female philopatric species, the biological mother's rate of interaction
with her infants is initially independent of their sex, but favors biological
daughters as they mature. Among Japanese macaques (*Macaca fuscata*),
for example, a young male interacts more with other males as he matures
and then eventually migrates to another troop (Nakamichi 1989), and so
"the mother-daughter grooming relationship persists through their lifetime,
whereas mother-son grooming terminates with male emigration" (Roney
and Maestripieri 2003: 183).

6　The rule also accounts for instances of Japanese macaque groups where the ascendency of the youngest biological daughter has not occurred. These seem to be cases with low rates of aggression between females, hence the rule is seldom made active and so a younger daughter has not learned a higher dominance ranking from her biological mother's support (Hill 1999).

7　Coalition formation (even if temporary), in which one female supports another engaged in an agonistic or competitive interaction with another female, has been suggested as the tradeoff for parasite removal through grooming, but this argument has not been without challenge (Henzi and Barrett 1999) and is contradicted by primate species where grooming occurs and coalitions are rare or do not form (see Figure 2.3).

8　Statistical analysis: $\bar{x}_{arboreal} = 2.29$, $n_{arboreal} = 7$, $s_{arboreal} = 0.07$; $\bar{x}_{terresrial} = 2.57$, $n_{terrestrial} = 12$, $s_{terrestrial} = 0.23$; $t = 3.72$, $df = 14$; $p = 0.001$, one-tailed test (unequal variances); neocortex data from Kudo and Dunbar 2001.

CHAPTER 3

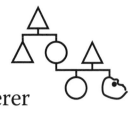

The Hunter-Gatherer Ending Point

We will use hunter-gatherer societies as our ending point. What distinguishes hunter-gatherer societies from the later horticultural, agricultural and pastoral societies that replaced most of them is a mode for food resource procurement that is not designed to positively affect the natural productivity of the resources being exploited. Unlike horticulture, agriculture, or animal husbandry where labor investment can increase the productivity of the resources being exploited, hunter-gatherer societies are dependent upon, and thus limited by, the natural productivity of resources. Individuals and groups within hunter-gatherer societies hunt, gather or fish according to the range, variety and abundance of resources as they occur in the environments that they exploit. Implements and tools may be designed to make the procurement of food resources more efficient and to reduce the risk of failure when hunting mobile prey (Torrence 1989; Read 2008b). Labor investment aimed at increasing the quality of implements and tools used in hunting, gathering and fishing may increase daily rates of return of food resources, but not their natural abundance. There are exceptions, such as the use of brush fires by Aborigines in Australia to increase the abundance of grasses as a way to attract grass-feeding marsupials. However, exceptions like these primarily imitate processes already present in nature, such as lightning-caused brush fires, and do not introduce innovative ways to disrupt and restructure the natural ecosystem as occurred later with plant and animal domestication. Domestication and the introduction of farming, gardening or pastoralism methods made possible increased food abundance through greater labor investment in the same physical area, enabling change in the organizational direction of societies with larger populations and densities.

For the most part, our evolutionary history as modern *Homo sapiens* is one of socially and culturally building on the cognitive properties already in place relatively early in our evolutionary history. Once modern-like hunter-gatherer societies were in place during the Upper Paleolithic, our evolutionary history has been primarily one of changes in the form of social organization, building on our existing biological and cognitive characteristics without requiring major change in those characteristics before new and novel forms of social organization would be feasible. There is general agreement that our ancestors by around 50,000 years BP were cognitively fully modern, and some push the date for cognitively modern *Homo sapiens* back to 70,000 BP or even earlier. Regardless of the precise date, we can characterize the hunter-gatherer endpoint of our odyssey using what we understand about the social and cultural properties of current or recent hunter-gatherer societies under the assumption that the biological bases for the cognitive capacities making these properties possible were already in place early in our history. Even the more recent forms of hunter-gatherer adaptations are within the range of the physical and cognitive capacities of our earlier hunter-gatherer ancestors.

This is not to say that the biological evolution of behaviors ended with the appearance of hunter-gatherer societies, but rather, with one or two notable exceptions, elaboration in the range and scope of human societies builds on biological properties already universal in our species: "The most profound changes in cultural evolution have occurred during the last 10,000 years.... No anthropologist today believes that there has been any appreciable change in 'human nature' during this time" (Carneiro 1981: 175–176). One of the more notable exceptions is the co-evolution that took place between lactose tolerance (a genetic trait) and the cultural practices that went along with the development of pastoral societies after the domestication of such animals as cattle, sheep, goats, camels, yaks and llamas, starting around 12,000 BP (Durham 1991).[1]

With domestication came the possibility of preempting for human consumption the milk produced by a female mammal for nursing her offspring. The default condition for adult humans is lactose intolerance due to the body ceasing to produce, by 4 years of age, the enzyme lactase that breaks down milk lactose. With animal husbandry, the meat and blood of domesticated animals became a controllable source of food. In addition, the milk of female animals when nursing their offspring was a potential food source. The potential was changed into actual through selection favoring both the cultural practice of making milk products—such as yoghurt, cheese and sour milk—in which lactose was partially broken down by bacteria and the biological selection of individuals whose bodies, for genetic reasons, continued to produce lactase. This co-evolution between cultural practice and a genetic trait—a possibility already envisaged by Herbert Spencer in his discussion regarding "the reciprocal influence of the society and its units" (1910: 11; as quoted in Carneiro 1981: 177)—eventually led to those modern populations in which individuals are milk-tolerant throughout their adult lives.

Simple Versus Complex Hunter-Gatherer Societies

Even with our attention restricted to hunter-gatherer societies, we still have a complex mix of societies ranging from small-scale, kin-based societies with a few hundred individuals to much larger, "complex" societies having forms of social organization usually associated with horticultural and agricultural modes of resource procurement. These complex hunter-gatherer societies evolved from simpler hunter-gatherer societies and are associated with relatively unusual ecological circumstances in which a few resources were highly abundant and amenable to extensive labor intensification, such as the salmon runs of northwest North America. These unusual circumstances made possible the formation of large-scale systems of social organizations based on the intensive exploitation of highly abundant resources. Because of the exceptional nature of these complex hunter-gatherer societies, we will further restrict our focus to just the kin-based, small-scale hunter-gatherer groups that exploited virtually every habitat on Planet Earth. Henceforth the term hunter-gatherer societies will be used only when considering kin-based, small-scale hunter-gatherer societies without any institutionalized political or administrative system.[2]

I will make no attempt to provide a complete survey of hunter-gatherer societies, but will focus instead on a few aspects central to the odyssey with which we are concerned. The aspects that concern us are those for which we have counterparts but not equivalence in the non-human primate societies from which our hunter-gatherer societies evolved. This will make it possible for us to frame our discussion of the odyssey leading to hunter-gatherer societies as analytically having a beginning and an endpoint. What I will show is that the endpoint is not simply an elaboration of trends we can trace back to the beginning point.

To make this argument, I will focus on four aspects. These are, going from more general to more specific, as follows. First, there is the conceptual boundary used by the members of one hunter-gatherer group to distinguish itself from other groups. Critical here will be a transition from boundaries between groups based on face-to-face interaction as occurs with the OW monkeys to boundaries defined through relationships among the members of a hunter-gatherer society.

Second, we will consider the mode of access to, and control over, food resources by the members of a hunter-gatherer society. The transition, as we will see, is from ownership determined by the fact of possession to collective ownership through which individual ownership then becomes a cultural construction.

Third is the sharing of certain resources among individuals in a hunter-gather society. It is well-known that sharing changes from rare and unlikely outside of female parenting among non-human primates to extensive, culturally determined patterns of sharing within hunter-gatherer societies. Sharing, as we will see, interfaces with the conceptual boundary of a hunter-gatherer society defined through cultural kinship criteria and derives from the collective ownership of resources. Sharing of a collectively owned resource takes place according to cultural rules— what is initially owned collectively may become individually owned and thereafter be shared according to individual interests and proclivities.

The fourth aspect is the egalitarian nature of hunter-gatherer societies (Boehm 1999). At a minimum, hunter-gatherer societies are egalitarian in that they lack a hierarchy based on institutionalized political positions. Individuals may temporarily take on leadership roles, but not institutionalized leadership positions. The egalitarian aspect goes beyond just the negative, the absence of institutionalized political positions, and includes the positive way members of hunter-gatherer societies see themselves regarding leadership positions. As one of the !Kung san (a hunter-gatherer group living on the edge of the Kalahari Desert in the northwestern part of Botswana in Africa) expressed it, "Of course we have headmen ... each of us is headman over himself" (Lee 1979: 457; quoted in Boehm 1999: 61).

Heterogeneity of Hunter-Gatherer Societies

For all four of these aspects of hunter-gatherer societies, culture plays a crucial role, leading to far greater heterogeneity in the social organization of hunter-gatherer societies within a single species than is the case for non-human primate societies. Hunter-gatherer societies have a diversity of forms of social organization that provide the social context within which individuals go about their day-to-day activities. Some groups, such as the Netsilik Inuit of the Arctic region, are organized vertically as extended families. The Netsilik refer to these extended families as *ilagiit nanagminariit* ('proper kin') (Balikci 1970). From a structural viewpoint, the *ilagiit nanagminariit* is based on father and son links, with daughters going to live with their husbands in their camps (a practice technically known as virilocal residence) upon marriage (though not all marriages are exogamous to the *ilagiit nanagminariit*). From a primate social organization viewpoint, the Netsilik Inuit are male philopatric. In contrast to the vertical organization of the Netsilik, the !Kung san have been organized horizontally into residence units structured through sibling and spouse-of-sibling links (Lee 1979). Spouse links are formed according to marriage rules that identify the cultural relatives with whom marriage cannot take place because such a marriage would, for them, be incestuous. The notion of philopatry as it occurs with non-human primate societies does not apply to them. Yet a different form of social organization is found in hunter-gatherer societies in Australia with complex (from our perspective) marriage rules expressed through the division of the society into named social groups having animal totems. These social groups are referred to by anthropologists as sections, and the Australian societies with a section form of social organization had 2, 4 or 8 sections. Marriage rules were also expressed through the section system by prescribing that a man of section X must marry a woman in section Y, and their child will be in section Z.

This variation in forms of social organization among hunter-gatherer societies does not have any obvious patterning regarding ecological conditions. Both the Kariera of Australia and the !Kung san, for example, live in comparable biomes. Both societies have a deep history that goes back thousands of years. Each has had time to work out effective modes of adaptation and both societies have worked out comparable ways to effectively use the plant and animal resources that are part of their environment. If the form of social organization relates to the ecological mode of adaptation, then we should find similar, certainly not strikingly different, forms of social organization in these two societies.

Although there is no single form of social organization that characterizes hunter-gatherer societies, their mode of food procurement nonetheless implies that hunter-gatherer societies, like our primate ancestors, have had a population density constrained by the natural availability of resources since their means of resource procurement do not increase the natural abundance of food resources through labor investment. The organization of individuals within the society into social units and the spatial distribution of these social units reflect the space and time dimensions employed as part of their adaptation to environmental conditions. For this reason, those aspects of hunter-gatherer societies constrained by demographic limitations will reflect aspects comparable to those found in non-human primate societies also constrained by the space and time dimensions of their environmental conditions.

Expansion of Hierarchical Levels: Widening of the Social Field

The organization of a single hunter-gatherer society through the division of the society into social units composed of families living together on a day-to-day basis resembles the division of a primate species into the social units we refer to as troops. Just as the OW monkeys are divided into troops of males and females, ranging in age from newborn to elderly, that interact with each other extensively and where the troop serves as the locus for individual behaviors on a day-to-day basis, hunter-gatherer societies have a similar division into residence groups that are the locus for day-to-day activities. The size of a residence group is comparable to that of many OW monkey troops—twenty to forty individuals with a mean of around thirty individuals across hunter-gatherer societies in widely differing ecological conditions. For the non-human primates, groups of this size appear to be a compromise between the pull for larger groups as a way to protect against predation and the push of group fission that may arise due to conflicts within the group stemming from factors such as competition within a group for access to food resources. In contrast, among hunter-gatherer societies the pull relates more to the advantages for food procurement, especially hunting of large animals, arising from living with groups composed of close kin and less to risks of predation or competition for access to food resources by the members of a social unit.

Although there is a similarity between the residence group and a troop, there is an additional hierarchical level in a hunter-gatherer society without a counterpart in non-human primate societies, namely the division of our species into societies, a division as real today with nation-states as it was when our species was divided into hunter-gatherer societies before the Neolithic revolution and the appearance of horticultural, agricultural and pastoral modes of food resource procurement. For the OW monkeys, as we have discussed in chapter 2, the troop is a more-or-less self-sufficient, closed social unit (except for transfer around the time of sexual maturity of one sex or the other to another troop for reproductive purposes) with a boundary expressed through antagonistic or avoidance behavior between troops. For hunter-gatherers, the society as a whole is the more or less self-sufficient, closed social unit. Like the relationship between two troops, the relationship between two hunter-gatherer societies is sometimes one of hostility or avoidance, thus suggesting an analogy between the hunter-gatherer society and a primate troop rather than between a residence group and a primate troop. However, regardless of whether we consider a hunter-gatherer society to be subdivided into residence groups that are the analog of primate troops, or *Homo sapiens* divided into societies as the analog of a primate species, the critical shift is from a two-level to a four-level hierarchy based on degree of inclusiveness and accompanied by an expansion of the social field. The social field for OW monkeys consists of the troop members and forms the base of their two-level hierarchy. For hunter-gatherer groups, the social field is the hunter-gatherer society as a whole and is located in the middle of a hierarchy for our species organized as a collection of hunter-gatherer societies. This shift in the hierarchical level for the social field leads to a form of social organization that is two orders more complex from a hierarchical point of view—and simultaneously an order of magnitude larger from a demographic point of view—than the troop structure.

Within a troop, as we have seen, females are linearly ordered in a dominance hierarchy structured by a simple rule that perpetuates the hierarchy. The rule also leads to a complex pattern for age relations among individuals within the linear dominance hierarchy (see Figure 2.4). Males are not part of the linear dominance hierarchy. In sharp contrast, residence groups in hunter-gatherer societies are composed of social units we refer to as families that typically include both a male and a female and her offspring. Families are hierarchically related to one another in the residence groups according to kinship generational differences among family members. Residence groups, in turn, are organized together in a nonhierarchical manner to form a hunter-gatherer society.

The hierarchical organization among families can crosscut residence groups, which brings up a second, and critical, difference between the organization of *Homo sapiens* into hunter-gatherer societies and the organization of a monkey species into troops. The relationships among individuals from different residence groups in a hunter-gatherer society are ones of social interaction among the members of different residence groups and not ones of antagonism or avoidance. Thus, while the social field in an OW monkey species is, for the most part, restricted to one's natal troop for the philopatric sex, the social field in *Homo sapiens* is the entire hunter-gatherer society.

The "How" Side of Adaptation: Implementing Traits

This brief sketch of some of the striking differences in structure between a monkey species and *Homo sapiens*, which will be addressed further in chapter 5, is insufficient as it does not address critical differences in how social organization in *Homo sapiens* is implemented in comparison to that in a non-human primate species. To see what is at issue, we need to consider the flip side of an adaptive argument that is often sidestepped or given insufficient attention. Arguments about adaptation generally focus on the selective advantage of behavior outcomes and often do not address the "how" side of an adaptation. An adaptive argument, though, has two aspects: the positive benefit that accrues from the adaptation and the means by which the adaptation can take place. The first aspect refers to the benefits or functionality that a trait may impart (whether physical or behavioral in form, individual or group in scope) that is relevant to whatever may be the salient selection mechanism. For the OW monkeys, there are Darwinian fitness benefits for individuals, such as protection against predators, arising from a troop form of organization. At the same time, for the troop organization to be maintained over time, the troop needs to minimize competitive conflict over resources between individuals within a troop since unchecked resource competition may disrupt the social coherency of a troop. The other side of the adaptive argument, then, relates to the means for implementing this organizational structure. Merely associating a benefit with a trait (where traits may include behaviors) does not ensure its appearance and implementation. The "how" side must also be developed: how are troops formed as a social unit in such a manner that the troop maintains coherency as a social unit over time?

Complicating the answer to the "how question" is the behavioral repertoire that may already be in place for the species in question, before or as part of the introduction of a new or modified form of social organization. Some species that form coherent social groups for protection against predation—such as schools of fish, flocks of birds and herds of ungulates—do so with limited social interaction among group members. For OW monkeys, we know that a behavioral repertoire evolved through which the troop structure is maintained by social interaction within a troop that focuses on interactions among matrilineally related females. Social interaction is a key aspect for the coherency of troops as social units, and so the answer to the "how" question regarding social organization for OW monkeys must also take into consideration that being highly social is part of their adaptation. In contrast, the kind of social unit we refer to as a herd or flock does not, by itself, require a high degree of sociality among group members. The "how question" was answered for the OW monkeys not through a herd or flock form of social organization but through incorporation of the sociality of primates in the form of linear dominance hierarchies based on a simple behavior mechanism implemented through social interaction among group members. At the same time, this mode of social organization made possible intensive social interaction, thereby leading to a feedback loop between mode of social organization and intensity of social interaction. Biological kin selection for the elaboration of social interaction behaviors among troop members became integral to the development of OW monkey forms of social organization, and the form of organization became integral to the implementation of biological kin selection. What then arose through selection was a link between social behavior and a linear dominance hierarchy implemented by a biological daughter entering the dominance hierarchy immediately below her biological mother. Let us think of this slotting of a biological daughter into the dominance hierarchy immediately below her biological mother as an *implementing trait* —it provides the means by which an adaptive feature, the dominance hierarchy, can be realized.

If, however, what might be an implementing trait for a potentially adaptive trait cannot be realized—that is, there currently is no feasible evolutionary pathway leading to an appropriate implementing trait— then a potentially adaptive trait will not be realized. Consider a social field that includes all the members of the different residence groups in a hunter-gatherer society as the adaptive trait. One of the adaptive advantages of the trait, in a general sense, lies in individuals being able to move, through social means, from a residence group in which resources are currently scarce to another residence group in which resources are abundant. Conditions under which this potentially adaptive advantage will be realized include those where the spatial scale for resource variability is much smaller than the spatial scale for the total area utilized by a society, so that some groups within the society will have surplus resources while others have a shortage of resources due to resource variability associated with the differences in the geographic location of the groups making up the society (Read and LeBlanc 2003). Conditions like this are not unique to hunter-gatherer societies. The adaptive advantage of socially integrated residence units applies to OW monkey species as much as it applies to hunter-gatherer societies.

So we can ask: why don't OW monkey species have an adaptation in which the social field includes several troops and is not circumscribed by the troop in which an individual is embedded? The reason lies in the answer to the how question: how does a species expand the social field to include between group social relations as well as within group social relations? We will examine the answer to this question in more detail in chapter 5 in the form of an implementing trait for hunter-gatherer groups. This will make it evident why the implementing trait was not possible within an OW monkey species. In this way, we will have identified a critical aspect of the transition from a primate-like form of social organization to a human-form of social organization.

Culture as an Implementing Trait

In our odyssey, we need to focus, then, on the question: by what mechanism(s) have hunter-gatherer societies expanded their social field to include the entire society with its subdivision into residence groups? The short answer is: culture. However, to say the answer is culture does not tell us very much since, depending on what we mean by culture in general and what is cultural about the enabling trait in particular, there may be no clear reason why the enabling trait is not within the reach of OW monkeys, or if not OW monkeys, the anthropoid apes, especially the chimpanzees.

Culture Traits as the Analogue of Genetic Traits

The term *culture*, at least when Darwinian evolution is used to model cultural evolution, has come to mean traits transmitted throughout a population by direct phenotypic transfer without depending upon genetic transmission. This follows from the widely quoted definition of culture attributed to the sociologist Edward B. Tylor, who lived in the last part of the nineteenth century. He defined culture as "that complex whole which includes knowledge, belief, art, morals, law, custom, and any other capabilities and habits acquired by man as a member of society" (Tylor 1924 [1871]:1). The phrase, "capabilities and habits acquired by man as a member of society," sets the stage for the more technical notion that culture has to do with direct, phenotypic transfer rather than with phenotypic transmission enabled indirectly through genetic transmission. With culture defined as traits transmitted through direct phenotypic transfer, culture is no longer a unique trait found only in *Homo sapiens;* evidently direct phenotypic transmission of behaviors occurs in other species. Perhaps one of the clearest examples is the one-handed grooming practiced by chimpanzees in the Mahale Reserve in central Africa discussed in chapter 1.

With the inclusion of direct phenotypic transfer as the mode of trait transmission from one individual to another, what constitutes culture and how culture may change has, for many researchers, come under the scope of the ideas that define Darwinian evolution. In analogy with biological traits, cultural evolution becomes simply change in the frequency of cultural, rather than biological, traits arising through the process of non-genetic trait transmittal. Selection is then defined through the process of phenotypic trait transmission acting over the members of a population rather than through reproduction. The mode of transmission, and hence the process by which selection takes place, will not be the same as for genetic traits, but the consequence, according to this argument, will be similar: evolution is measured as change in the frequency of a trait over a population. Of course, there must also be a means by which the traits being transmitted came into existence initially, and this, as with biological traits, can occur through imperfect transmission.

In this Darwinian framework, the enterprise of understanding human societies appears to come under a single, overarching explanatory framework that does not incorporate an absolute distinction between nature and nurture; that is, between biological and cultural processes. Both nature and nurture become two sides of the same evolutionary phenomenon, namely change in the frequency of traits in a population arising over time through the process of trait transmission and selection.

Culture as a "Complex Whole"

As appealing as this framework may appear at first glance, the reduction of culture to the mode of transmission is too simplistic. That culture, however we define it, involves transmission in a social context directly from one phenotype to another goes without question, for how else do we become encultured into the cultural milieu in which we are born and raised? The mode of transmission is not the defining characteristic we want, but what constitutes culture beyond the truism that the mode of culture transmission is through direct phenotypic transmission. The mode of transmission simply makes explicit what we already understand: culture is not just another kind of biologically based behavior. Rather than saying what it is not, we want to understand culture from the viewpoint of questions such as: what is there about culture that enables an expanded social field that includes not only those with whom one interacts on a day-to-day basis, but others with whom interaction may only be occasional and even others with whom no prior interaction has taken place? To answer such questions for our hunter-gatherer societies, we need to return to the first part of Tylor's definition: "that complex whole." To understand culture as a complex whole, we need to see how culture in this sense involves more than just the non-biological transmission of traits. Consider the comment an Inuit spokesperson made about his culture. The comment makes evident what Tylor meant by a "complex whole": "In Nunavut, we call our culture Inuit Qaujimajatuqangit [IQ].... IQ refers to a way of viewing the world and the values that go along with living a proper life. It is an approach that defines Inuit. It involves many aspects, including strong values related to social harmony, mutual sharing and assistance, and honesty" (Okalik 2001: 13). Culture, from an Inuit perspective, cannot be reduced to the frequency of cultural traits as it is not individual traits, per se, that determine Inuit culture, but a "complex whole" of values and relations that structures and organizes the Inuit way of life in all of its many dimensions.

What we want, then, is to move away from the simple notion of culture defined through mode of transmission, yet to keep our discussion of an enabling trait for hunter-gatherer societies anchored, historically, in the Darwinian evolutionary foundation from which a cultural framework has arisen. For how ever we may understand culture as it has developed in human societies, its evolutionary beginnings in non-human primate societies must be understood in Darwinian terms. To see the issue more clearly, we need to reexamine what constitutes the population for measuring Darwinian evolution through change in the frequency of traits. This will make evident what we need to focus on in the transition to hunter-gatherer forms of social organization.

Biological Population: Reproduction Within a Species as an Information Boundary

For biological traits transmitted through sexual reproduction, the population over which change of trait frequency is a relevant measure of evolution is a species defined by male and female individuals capable of viable reproduction from one generation to the next. The definition can sometimes be fuzzy in application because Mother Nature is not so kind as to always provide us with biological systems having simple boundaries. The definition of the boundary of a species through viable reproduction has exceptions, but on the whole the idea behind the definition stems from the functionality of a species boundary from an evolutionary viewpoint. The species boundary serves the function of ensuring that new combinations of genetic materials lead to individuals who are still within a workable range of variation for the current mode of adaptation of a species to its environment. Sexual reproduction implies that the traits of offspring will always be a compromise between the genetic material provided by a female and a male. With qualitative traits such as eye color, the compromise is generally despotic or "all or nothing." For quantitative traits such as stature, the compromise tends to be more democratic, with a trait value falling between those of the biological mother and the biological father. As a consequence, if there were cross-species breeding for quantitative traits, then the offspring could have trait values less fit than the trait value of either parent because of having a morphological form that is suited for the biological adaptation represented neither by the species of the biological father nor by the species of the biological mother. For example, if one could introduce cross-breeding on the Galapagos Islands between a finch species with long, thin beaks that work well for retrieving soft food sources such as insects through narrow constrictions and a finch species with short, heavy beaks that work well for breaking open seeds with hard shells, the offspring would have beaks somewhere between these two forms. As a result, the offspring would have a trait that is not effective for either food source, hence the offspring of interbreeding finches would be less fit than the offspring of finches that did not interbreed. Under these conditions, there will be selection for any trait that reduces the likelihood of interbreeding, or ensures the non-viability of a fertilized egg (thereby avoiding the fitness cost of investing in offspring with low fitness) when there is interbreeding. Thus, there will, in general, be natural selection for boundaries protecting against interbreeding. These boundaries have the function of limiting the range and mix of genetic information that can be used to form new individuals.

A species boundary in this sense is relevant not only for morphological traits, but for traits relating to social interaction as well. This notion of a trait boundary connects the forms of coherent social organization of hunter-gatherer societies back to their biological foundations. For social coherency to arise, social interaction depends upon the coordination of behaviors by the individuals involved. Coordination makes it possible for one individual to act in anticipation of the action of another in response to one's own actions, whether or not there is conscious awareness of the meaning of the actions. In the OW monkeys, coordination of females arises through the dominance hierarchy since a dominant female can take action towards a lower ranking female under her (implicit) assumption that the lower ranking female will be submissive when confronted by a higher ranking female; lower ranking females have "learned" to be submissive when interacting with dominant females. When this coordination between the higher and lower ranking females does not take place, conflicts may arise, such as when the actions of one female are taken in accord with her interests but without coordination with the interests of other females. Competition over food resources can arise when there is lack of coordination over access to resources and, absent coordination, being a member of a group may be detrimental for individual fitness when group members compete over resources. More generally, absent coordination among group members, there is little fitness benefit to be accrued by an individual for being part of a social group. For biologically based behaviors, maintaining behavioral coordination relates to species boundaries in the same way as it does for morphological traits.

Cultural Population: Kinship Based Social Interaction as an Information Boundary

For hunter-gatherer groups, we do not find a boundary determined biologically through reproduction, but conceptually through a group's self identification. The !Kung san define themselves, as we noted in chapter 1, as the *ju hwansi*—"the real people." Who are the real people? They are their kin from a cultural kin viewpoint. Those who are not kin are strangers, and they refer to strangers as *ju dole*, where *ju* means people and *dole* means harmful or dangerous. So someone who is not their kin is considered to be harmful or dangerous. In some hunter-gatherer groups, this perception of a stranger translated into a simple expediency: kill a stranger/non-kin before he has a chance to harm you. The Aboriginal hunter-gatherer groups in Australia in traditional times before the incursion of outside groups considered strangers to be "deadly enemies" (Curr 1886 [2005]: 64). For these and other hunter-gatherer groups, evidently being a non-kin meant being outside the "real people" among whom social life and social interaction and mutual aid and assistance take place. We can see this in the Netsilik Inuit story (Balikci 1970) about how the orphan girl, Nuliajuk, became goddess of the seals after she drowned when pushed off a crowded raft. The story makes the point that she had no kin on the raft to help and protect her; that is, one can only count on one's kin to come to one's aid and provide help. The cultural boundary, then, is between those individuals who are part of the same moral world and those who are not; between those whose behavior is predictable and those whose likely behavior is unknown (or at least believed to be unknown); between those whose behavior is restrained by expected behavior associated with proper kin relations and those with whom behavior is not constrained by proper kin relations.

Like the species boundary that relates to the maintenance of coordination among genetically based behaviors, the cultural boundary relates to the maintenance of coordination among individuals for culturally based behaviors. Individuals enculturated within the same cultural milieu understand the meaning of the behaviors of others in a similar way, and others respond to an individual's behavior in a similar manner through shared, cultural meaning of actions. We act in a cultural context in anticipation of how others will act, and our common cultural background makes coordination more likely across different individuals, even if it occurs non-consciously. For someone in a hunter-gatherer society, killing an animal or foraging for plant foods is not an isolated, individual act but one that has cultural meaning and, through that meaning, expectations are generated in others and these expectations are understood in a similar manner by all the members of the society. Is the animal that has been killed to be divided and shared with others, and if so, with whom and by what criteria? What about plant foods or other resources? Are they shared, and if so by whom and under what conditions? These are questions that have cultural answers, and because they have cultural answers, the outcomes regarding who has rights of access to those food resources and under what conditions (or at least the framework within which these outcomes are worked out) are known in advance, at least in a general way, to those sharing that culture. In this way, individual interests may become subordinated to group interests through culturally constructed expectations, and through that subordination the actions and behaviors of others can be anticipated. And when anticipations are not met, those actions and behaviors can be judged and sanctions imposed to the degree it is possible to impose sanctions by the persons involved. Hunter-gatherer societies are without political systems and institutions designed to enforce conformity of behavior to community norms and values, so sanctions may simply be social ostracization by one's kin, a very powerful form of sanction in societies where one's survival depends on coordination of behaviors among those recognized as kin.

The distinction between kin and non-kin in human societies, or what has been expressed as the "sentimental attachments that distinguish kin from nonkin" (Brown 1991:105), highlights the shift in the evolution of human societies to behaviors based on expectations—expectations about the behavior of others that are derived from the relation or role that another has to oneself without first requiring prior interaction with that individual to determine how another individual is likely to behave. We expect cultural kin to be supportive simply because of being cultural kin. Unlike biological kin, a relationship that provides the necessary (and limiting) conditions on individuals under which there may be selection for biologically based social behaviors, cultural kinship in human societies is biologically open-ended in that it directly invokes expectations about behaviors associated with culturally recognized kinship relations separate from the degree of biological relationships among the individuals involved. The expectations are reciprocal: the behaviors expected from those recognized as one's cultural kin and one's understanding of the behaviors one should engage in toward those recognized as one's cultural kin. Fortes expressed it succinctly in his "axiom of amity"; that is, kinship entails *prescriptive* altruism exhibited in the ethic of generosity" (1969: 232; emphasis added); i.e., those who are culturally recognized as kin to each other are expected and assumed to be altruistic in their behavior simply by virtue of the fact of being kin to one another. In the context of hunter-gatherers, we see it in expressions such as "the moral universe of the Mardu [a hunter-gatherer group in western Australia] is populated solely with relatives" (Tonkinson 1991: 57, as quoted in Dousset 2005: 22). Culturally constructed kinship in hunter-gatherer societies is not simply a means for expressing the degree of relationship among individuals, but frames the actions that make up day-to-day living by forming a bounded, social universe composed of one's kin.

Kinship Conceptual Basis for a Social Boundary

With a culturally defined social boundary expressed through the persons who are kin to one another, we need to understand the conceptual basis for someone being recognized as the kin of someone else. Kinship relations as they occur in human societies are ultimately grounded in both biology (through reproduction) and culture (through marriage, a universal, though highly variable, institution in its implementation). However, knowing the ultimate grounding does not inform us as to the particulars of a cultural kinship system, especially the kinship terminology that defines and structures the kin relations the members of a hunter-gatherer society have to each other. In all societies, and not just hunter-gatherer societies, societal members have, as part of their enculturation, cultural knowledge regarding the kin terms used to both define and refer to those to whom one is related through kinship. Anthropologists call this set of kin terms a kinship terminology. For English speakers, our kin terms include father, mother, brother, sister, son, daughter, grandfather, grandmother, uncle, aunt, nephew, niece, and cousin, among other terms, as well as terms such as father-in-law, mother-in-law, among others, that express kinship relations constructed through marriage.

Kin terms have generally been considered to be a means for labeling categories of persons in accordance with their genealogical relation to the speaker traced through parent and child relations, such as the English terms *brother* or *sister* used to refer to anyone with the genealogical relation parent's child to oneself, according to sex, or *aunt* being the term used by ego to refer to anyone with any one of the genealogical relations, mother's sister, father's sister, mother's brother's wife or father's brother's wife, to ego. While it is true that kinship terms have this function, more important for our purposes here is the way in which kinship terms define, structure and provide the conceptual organization for those we construe to be our kin. Whereas genealogical relations express the connection of one person to another through chains of parent and child links (or, more technically, through *filiation*), kin terms express the conceptual link between one person and another by reference to a constructed, conceptual system of kinship relations. The system of kin relations makes it possible to compute kin relations using the kin terms, much like we do arithmetic computations with numbers. This means it is possible to compute who is related to whom and in what way by reference to the cultural knowledge embedded in a kinship terminology by the conceptual relations expressed through the kin terms.

Without a kin term computational system, a person would need to keep track of hundreds or thousands of genealogical pathways to know all the possible relations between individuals in a group even as small as a hunter-gatherer society of several hundred persons. With a kinship terminology system, one only needs to know a dozen or so kin terms and the conceptual relations among these terms from which the kin term relationships between two individuals may be computed by reference to the kin term relations each has to a third person. The Kariera, a hunter-gatherer group in western Australia, compute kin relations directly in this manner from kin terms: "The method of determining the relationship of two individuals is extremely simple. Let us suppose ... that two men, A and B, meet each other for the first time. The man A has relative C who is his *mama*. At the same time, C is the *kaga* of B. It immediately follows that A and B are *kumbali* to each other" (Radcliffe-Brown 1913:150–151, italics mine). Radcliffe-Brown is stating that the Kariera can compute the kin term relationship of A to B (and reciprocally, the kin term relationship of B to A) just by reference to their kinship terminology and without any information about the genealogical relations among A, B and C. The computation, of course, cannot be inferred by us from their kin terms taken in isolation since *mama* ('father'), *kaga* ('maternal uncle') and *kumbali* ('man's male cross-cousin') are not our kin terms. Instead, the computation expresses cultural knowledge the Kariera have about their kinship terminology and the conceptual relations among their kin terms. We, as English speakers, make a similar kind of kin term computation with our kin terms when we know that "cousin" is the kin term we would use to refer to person B if we (properly) refer to person A as aunt and person A (properly) refers to person B as daughter or son; that is, we can express the relations among these kin terms using the kin term products, *daughter of aunt* or *son of aunt*. We, as users of our terminology, know that these products define the relational meaning of our kin term cousin by reference to the kin terms aunt, son and daughter. This is similar to our number system in which we have rules expressing how counting numbers may be combined and equated with other counting numbers (which we may also express symbolically), such as two objects plus four objects yields six objects (or symbolically, $2 + 4 = 6$). The computational aspect of kinship terminologies underscores the way kinship in human societies is a cultural construct and neither simply a reflection of biological kin relations nor of behaviors. Whereas biological kin are independent of cultural context, the system of computational rules for a kinship terminology is culture specific and varies in form from one society to another.

To keep the biological and cultural notions of kinship distinct, we will refer to the system of relations expressed through a kinship terminology as *cultural kin,* in contrast with *biological kin* determined by biological reproduction. Those who are our cultural kin will overlap with those who are our biological kin, for it is evident in our terminology that kin terms such as parent, grandparent, great grandparent, etc., make distinctions paralleling the actual degree of genetic relatedness for the persons culturally related according to these terms. Other terms such as cousin, though, refer to persons in different generations, hence refer equally to persons with different genetic distance. Even more extreme, marriage creates for us cultural kin relations between persons who are not biological kin. In our culture we consider sexual relations between close biological kin to be incestuous. Correspondingly, we have cultural rules in the form of incest taboos that prohibit marriage between close cultural kin. Husband and wife, in our culture, are typically biologically unrelated yet become close cultural kin through marriage. Similarly, we consider adoption to be an alternative way to establish cultural kin relations, and adoption typically takes place without genetic relatedness between the person being adopted and the adopting couple.

The rules and structural equations for the computational logic of a particular terminology are cultural, meaning that they are not derived from biological properties. Instead, these rules and equations define kinship relations as a logically structured system of culturally determined kin term concepts. The connections among the concepts can be expressed through the kin term products computed by individuals to determine their respective kin relationships. With the natural numbers and multiplication, product rules are given systematically as a multiplication table for the basic numbers from 1 to 9, from which we can read off rules such as 2 x 3 = 6. For kin terms, the rules for kin term products such as *daughter* of *aunt* is *cousin* (or in equation form *daughter* o *aunt* = *cousin* by using 'o' to stand for "of") can be presented systematically and visually in what we will call a kin term map. The kin term map expresses the cultural knowledge individuals have about the interconnections among the kin terms in their kinship terminology, and this knowledge is called upon when computing kinship relations. Kin term maps for kinship terminologies provide us with a simple, visual way to see crucial structural differences among kinship terminologies. We will now use kin term maps to indicate how different kinship terminologies can be for hunter-gatherer groups, even for groups that have comparable ecological adaptations. To do this we, first need to explain in more detail what is meant by a kin term map.

Kin Term Maps

Cultural knowledge about kin terms can be elicited experimentally (Leaf 2006; Leaf and Read forthcoming) through systematically asking questions of the form: "If you (properly) refer to a person by the kin term K and that person (properly) refers to another person by the kin term L, then what is the kin term M that you use to refer to that last person?" where the kin term L is one of the terms, such as father, mother, son, daughter, husband or wife (for English speakers), used to express one of the positions in what we can call a family space. Questions like this will elicit a new kin term not yet elicited, elicit a previously elicited kin term, or elicit no kin term. For example, if the person making the enquiries currently only knows about the English kin terms father and mother and asks, "If you (properly) refer to someone by the kin term father and that person (properly) refers to another person by the kin term mother, then what is the kin term you use to refer to that other person?" the answer will be a new kin term (for the enquiring person), in this case *grandmother*. If one already knows about the kin terms brother and cousin, asking about brother of cousin will elicit the reply, *cousin*, and not a new kin term. Lastly, if asking about father of father-in-law, the reply will be that there is no kin term corresponding to this product. In this way, the elicitation procedure leads not only to the kin terms making up the kinship terminology, but to the conceptual boundaries for the system of kin terms.

Kin Term Map for the American Kinship Terminology

We now visually illustrate the outcome of systematically eliciting kin terms in this manner through a kin term map for the American kinship terminology (AKT), a terminology probably familiar to most readers of this book (see Figure 3.1). Each kin term corresponds to a node in the graph. An arrow, whose form (shape of arrowhead and solid or dashed line for the shaft) corresponds to one of the terms for the positions in the family space, is drawn from one term, K, to another term, L, when the kin term product of the kin term corresponding to the arrow type with the kin term, K, yields the kin term L. Thus an arrow corresponding to the term *son* points from *mother* to *brother* since the kin term product *son* of *mother* yields the kin term *brother* for users of the AKT.

The kin term map makes visually evident structural properties of the AKT, such as the lineal, ladder-like, pattern of terms extending upwards and downwards from the "self" position and a similar pattern for the collateral terms extending downward from each of the ascending, lineal terms. We can use the kin term map to compute products of kin terms and to determine the kinship relation between one person and another. For example, suppose you are at a wedding and your grandmother points to a man across the room whom you have not met before, and she refers to him as her nephew. From her comment you

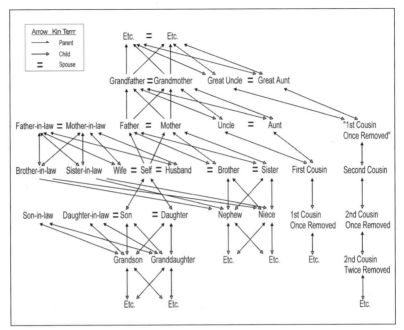

Figure 3.1 Kin term map of the American/English kinship terminology.

know that he is related to you, but how? From the kin term map, the kin term product *son* of *son* of *mother* starts at *self* and ends at the kin term *nephew*, so the product (*son* of *son* of *mother*) of *grandmother* will start at *grandmother* and end at the kin term relation that her nephew (that is, *son* of *son* of *mother*) has to you. It may be easily seen in Figure 3.1 that (*son* of *son* of *mother*) of *grandmother* is *first cousin once-removed*, so a person who is your grandmother's nephew is your first cousin once-removed according to the AKT.

As we can see from this example, the computational logic for computing kin relations is inherent in the kinship terminology and expresses the kinship cultural knowledge embedded therein. In the AKT, part of that embedded knowledge is expressed through the unending vertical chains of kin term products corresponding to the cultural idea for English speakers that no matter how distant an ancestor may be, that ancestor has a kin term relation to speaker, hence is recognized formally as one of speaker's kin.

The kin term map is a straightforward way to identify differences in cultural knowledge and kinship concepts in different societies. With this in mind, now let us turn to the kin term maps for the !Kung san and the

Kariera kinship terminologies to see the striking differences in cultural knowledge embedded in kinship terminologies even within small-scale hunter-gatherer societies having similar ecological adaptations.

Kin Term Map for the !Kung San Kinship Terminology

The kin term map for the !Kung san terminology is shown in Figure 3.2. That there are extensive structural differences between this terminology and the AKT is immediately obvious. Perhaps the most notable difference is the presence of two substructures in the !Kung san terminology.

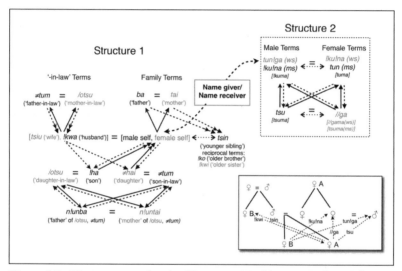

Figure 3.2 Kin term map for the !Kung san kinship terminology. The terminology consists of two disconnected structures. Structure 1 consists of the kin terms for immediate family and spouse relations. Structure 2 is a simple structure made up of a sex/spouse horizontal dimension and a vertical parent/child dimension. The two structures are conceptually connected through the name giver/name receiver relationship between a newborn child and the person for whom the child is named. The implications of the name giver/name receiver relation are illustrated in the diagram at the lower right of the figure in which the dashed arrows identify the kin relation between two persons. In this diagram, a female at bottom right received her name, A, from her maternal grandmother. She refers to her mother's sister and her mother's sister's husband by the +1, -1 generation terms //ga and tsu, respectively (from Structure 2 computed with respect to her name giver). Her sister received her name, B, from their father's sister and so she refers to the same persons with the 0 generation terms !ku!na and tun!ga, respectively (also from Structure 2 computed with respect to her name giver). The woman named A refers to her sister's name giver by an older sibling term !kwi and reciprocally is referred to by that person with a younger sibling term tsin.

The AKT has no counterpart to these two substructures. The first of these two substructures expresses the family and marriage kin relations around speaker, including a relative age distinction for the sibling terms for which a sex difference only occurs in the terms for 'older sibling.' The second structure is based on four positions connected vertically using kin term products and has the pairs of sex distinguished terms *ba* ('father')/*tai* ('mother') and *!ha* ('son')/*≠tum* ('daughter'), where the symbols !, // and ≠ refer to click sounds in the Khoisan language spoken by the !Kung san. These pairs of terms are connected horizontally with products using sibling and spouse terms. This second structure has terms without genealogical definition since the terms in the four positions in this structure are not connected by kin products to the self position in the first structure. Instead, the two structures are linked through a name giver/name relation that is fundamental to identifying and structuring the domain of kin among the !Kung san (Marshall 1976).

The name giver/name receiver relation is established when the parents give the name of a close relative to a newborn. A first-born male is usually named for the paternal grandfather (father's father), the second-born male is generally named for the maternal grandfather, and other males are generally named for relatives who are either in the first or second generation above speaker. A similar pattern holds for newborn females, but with greater variability in naming a first or second born daughter. The first-born daughter is generally named for mother's mother, and the second-born is often named for the father's mother.

For both boys and girls, naming establishes a conceptual identity between the name receiver (the newborn) and the name giver (the person whose name was given to the newborn). This conceptual identity between name giver and name receiver establishes the link between the two structures.

The conceptual identity means that a newborn person will calculate her or his kin relations outside of the family space as if she or he were her or his name giver. Thus, as shown in the inset in Figure 3.2, a female given the name A of her maternal grandmother will refer to a daughter of her maternal grandmother (her mother's sister) by the kin term, *//ga,* and to the husband of this woman by the kin term, *tsu,* since both of these persons are one generation below the maternal grandmother. However, if the sister of A received her name B from her father's sister, then she will refer to these two persons by the kin terms *!ku!na* and *tun!ga,* respectively, since they are on the same generation as her name giver. The logic of the name giver/name receiver identity is also carried over to the name giver of speaker's sibling; speaker will refer to a sibling's name giver by a sibling term, and reciprocally the name giver will refer to speaker by the reciprocal sibling term.

From our perspective, the terminology violates our expectation that if persons X and Y have the same genealogical relation to a third person, then each of X and Y will refer to that person with the same kin term. However, this just means that neither biological nor genealogical relations alone are the driver for the kin relations expressed in the !Kung san kinship terminology. Instead, kin are identified and structured through

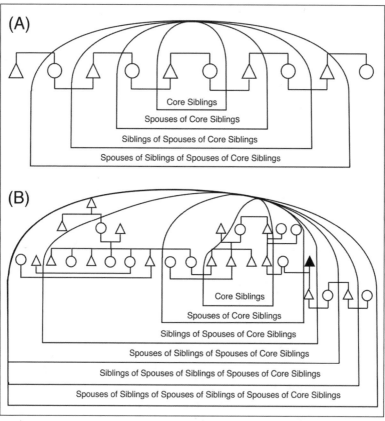

Figure 3.3 (A) Schematic diagram showing the structure of a !Kung san residence group. Triangles are males and circles are females. Solid triangle is a deceased male. A horizontal line above and connecting symbols shows sibling relationships and a horizontal line below and connecting a pair of symbols shows a marital relationship. (B) Genealogical diagram for the members of Dobe residence group in 1968 that matches the schematic diagram. *Redrawn from Lee 1979: Figures 3.3 and 3.6, by permission of publisher.*

the logic of concepts such as the name giver/name receiver relationship coupled with a division of kin terms into one structure based on positions in a family space and a second, minimal structure for incorporating vertical (parent/child) and horizontal (sibling and marriage) relations.

The concepts embedded in the second structure also provide a model for the membership and structure of residence groups. Horizontally, the kin term structure implies a sequence of sibling and spouse links with attached vertical, parent and child links (see Figure 3.3A). The residence groups of the !Kung san are structured in precisely this manner. A residence group has a core set of siblings, spouses of those siblings, siblings of the spouses of those siblings, and so on, including parent and child relations (see Figure 3.3B). The core set of siblings are the *k"ausi* (owners) of a waterhole, and visitors to a waterhole must obtain permission from them for access to food resources in the *n!ore*— the resource area around the waterhole— and others are linked to them horizontally in the manner expressed through the sibling and spouse links in the second structure.

Kin Term Map for the Kariera Kinship Terminology

We now compare the kin term map for the !Kung san terminology with the kin term map for the terminology of the Kariera from Australia (see Figure 3.4). The structure of their kinship terminology is as different from the !Kung san terminology (and from the AKT) as the !Kung san terminology is from the AKT. Name relations are not involved. Instead of two, connected structures, the kinship terminology has a single, symmetrical structure in which there are four "lines" of kin terms connected by taking products of a kin term with the term used to refer to one's son or to one's daughter. One of the striking aspects of the Kariera terminology, and the subject of extensive theorizing about the relative importance of vertical parent/child links versus horizontal marriage (affinal) links in social organization, is the use of the kin term *ñuba* ('cross-cousin') for spouse of speaker since the kin term *ñuba*, for a male speaker, will be one of the kin term products (a) *kundal* ('daughter') of *kaja* ('ascending brother') of *nganga* ('mother') or (b) *kundal* ('daughter') of *turdu* ('ascending sister') of *mama* ('father'); hence, the logic of the terminology specifies the criteria for deciding on a spouse for a male and, reciprocally, the spouse for a female. That is, a spouse must be someone referred to by the kin term *ñuba*. This prescription follows from the symmetrical structure with its "lines" of vertically connected kin terms. The symmetrical pattern will be preserved only if 'child' of ñuba is also 'child' of speaker, and that requires ñuba to be spouse of speaker for consistency with the kin term products. Thus, the structural logic of the terminology requires a marriage rule specifying that speaker must have a spouse from the persons referred to by speaker as ñuba. As a consequence, there are no affinal kin term positions separate from the consanguineal kin term positions.

Another major difference arises from the fact that the kin term *mama* ('father') refers not only to genealogical father, but to genealogical father's father's son, father's father's father's son's son, and so on. A similar comment applies to the kin term *nganga* ('mother'). The lack of a kin term for genealogical father alone (as is the case in the AKT) and the inclusion of the above list of genealogical relations under the *mama* ('father') term (and a similar list of genealogical relations under the *nganga* ['mother'] term) stems from the centrality of the sibling relation (Radcliffe-Brown 1950) in identifying and providing structure for the kin of speaker (Read, Fischer, and Chit Hlaing forthcoming).

The centrality of a sibling relation in the structuring of kinship relations can be seen in other groups with a similar pattern for the 'father' and 'mother' kin terms. For example, among the Tangu (a horticultural group in New Guinea), "siblingship is the determinant that descent [parent-child links] might have been expected to be … descent was probably always calculated from siblingship … and siblingship rather than descent always provided the definitive norms of social behavior" (Burridge 1959/60: 128, 130).

The pattern given above for those referred to by the kin term *mama* ('father') or *nganga* ('mother') in the Kariera terminology follows from the logical implications of having a sibling relationship be one of the generating concepts for the terminology (see Leaf and Read forthcoming; Read and Behrens 1990). Additionally, the structure of the Kariera terminology, though ego-centric in its form, is logically consistent with a division of the society into four, named socio-centric groups (Leaf and Read forthcoming) that anthropologists refer to as sections. The section system provides the Kariera with an alternative, socio-centric way to specify who can marry whom.

Kin Term Map: Structure and Social Organization

These differences in the kin term maps are reflected in the differences in their respective forms of social organization. The !Kung san, as discussed above, have local residence groups based on alternating and repeated sibling and spouse relations linking the members of the residence group.

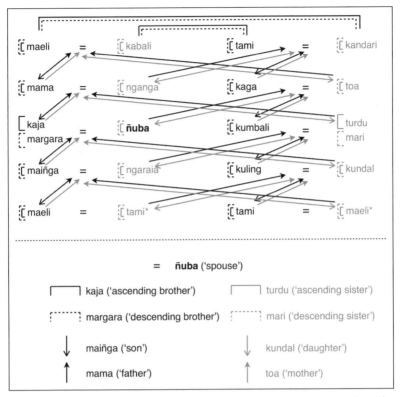

Figure 3.4 Kariera kin term map from the perspective of a male speaker. The vertical arrows point to the "=" sign only for clarity of the diagram and should be understood as pointing to the kin term matching the sex marking of the arrow; e.g., *mama* of *mama* is *maeli*. The vertical sibling symbols show that same-sex kin term products with sibling terms are reflexive; e.g., *kaja* of *mama* = *mama* = *margara* of *mama*. The horizontal sibling symbols refer to a cross-sex kin term product with a sibling kin term; e.g., *turdu* of *maeli* is *kandari*, and *kaja* of *kandari* is *maeli*. Black text—male marked terms; gray text—female marked terms; bold text—neutral terms. *Modified from Radcliffe-Brown 1913: Table 1.*

The only more inclusive social group is the !Kung san society as a whole demarcated by mutually recognizable kin. In contrast, the Kariera have residence groups based on father/child links that are integrated globally through their four named sections that are linked vertically by parent/child ties and horizontally by *ñuba* marriage relations.

These differences in the two terminologies do not correspond to behavioral differences related to ecological adaptation. Both groups are in arid environments in which resources are unevenly distributed in any given year and with changing abundance across years due to variability in rainfall. In both cases, spatial mobility by residence groups and by families temporarily moving between residence groups is crucial to their respective ecological adaptations. In both cases, family mobility is based on cultural kin relations between a family in one residence group and relatives in other residence groups. Instead of ecological adaptation being determinative of the structure of the kinship terminology, the behaviors that make for effective ecological adaptations are worked out according to their system of kin relations expressed through a kinship terminology.

That a kinship terminology is not determined directly by ecological adaptation can also be seen in Central Arctic Inuit groups such as the Netsilik that required groups of a size—when hunting seals through the pack ice during the winter—that could not be formed just with culturally close relatives. Seals were hunted in the winter when they came up for air in their breathing holes maintained in the pack ice. There was a strong chance element in killing seals due to the absence of any surface clues as to which breathing hole would be used by a seal when it needed air. To successfully hunt seals when they came up for air in the breathing holes, the Netsilik needed winter camps of sixty to one hundred persons so as to have adult males stationed at enough breathing holes to ensure that the hunters collectively obtained an adequate supply of seals during the winter. The families living together had to work cooperatively. From a cultural kinship viewpoint, cooperation is problematic with distant cultural kin, not close cultural kin. A winter camp of one hundred persons could not be formed from close kin. Kin networks were restricted in size by cousin marriages (Balikci 1970). Their solution to assembling the number of close cultural kin required was to form a system of sealing partners, called *niqaiturvigiit* (Damas 1972: 227), among the hunters. A seal killed by a hunter was shared with his *niqaiturvigiit* (sealing partners).

Each of the sealing partners for a hunter corresponded to the parts of a seal determined by a standard division of a seal into 12 parts plus 2 additional parts, one for the hunter and the other for his children (Van de Velde 1956; Balikci 1970). The parts of the seal were given by the wife of the hunter to the wife of the sealing partner corresponding to that part of a seal. Thus if a man were the "right front flipper" partner of the hunter, he would receive the right front flipper of a seal killed by his partner. The sealing partners were either selected by a man's father or mother at his birth or the partnership was passed on from his father to him. The males selected as sealing partners for a son often had a distant cultural kin relation to him (Damas 1972). In this way, the sealing partners complemented kinship relations and made it possible for otherwise distantly related males to be linked in a kinship-like manner through a system of reciprocal sealing partner relations. Since the sealing camps in the winter time needed to be large and distant cultural kin had to be included, the system of sealing partners made partners of males and provided rules of cooperation and sharing for individuals who otherwise would only have had a distant cultural kin relationship to each other.

Ownership and Sharing of Resources

We now consider the rights that individuals, families and residence groups have to food resources in hunter-gatherer societies. Our interest here is less with the means by which resources are obtained than the rights that are recognized by being a member of a residence group. As regards the means for procuring resources, extensive research, using a variety of modeling approaches, including optimal foraging, costly signaling, utility maximization, and the like, leads to essentially the same general conclusion: hunter-gatherer societies have worked out effective means for the procurement and distribution of resources, taking into account the implements needed to obtain resources, the size of the labor pool involved in the procurement of resources, modes of dealing with risk, fitness benefits, spatial location of residence groups, and so on. This research, focusing on the *facts* of resource procurement, confirms in a general way the effectiveness of hunter-gatherers in procuring resources from their environment. However, usually left unstated in this research are the rights and obligations of individuals, and of individuals as family and group members, to resources through cultural rules of ownership.

Typically, resources in hunter-gatherer societies are collectively owned at the level of a residence group in that all members of a residence group have rights of access to the resources in the spatial region recognized both by the members of the residence group and by other residence groups as belonging to, or associated with, that residence group. Collective ownership of resources by the residence group members implies that these resources are not open to access by members of other residence groups. Collective ownership in this sense is functionally analogous to a region exploited for resources by a primate troop with a defended boundary, but with the critical difference that in hunter-gatherer societies those outside of a residence group may have access to resources under the ownership of that group through cultural kin relations.

Access to resources "owned" by a different residence group than one's own is typically obtained through establishing temporary residence through one's cultural kin who are members of the resource-owning residence group. Cultural kinship plays a key role through both providing the idiom by which the criteria for membership in a residence group is expressed and the means by which permission may be obtained to forage for resources associated with a residence group.

What is collectively owned implies shared access; that is, each person in the residence group has right of access to resources in the area associated with that residence group. Collective ownership contrasts conceptually with individual ownership, and individual ownership may be established through the labor and actions of individuals expended in procuring resources. Though superficially similar to the individual ownership through possession in non-human primate societies, individual ownership is part of the cultural milieu of rights and obligations associated with being a member of those considered to be "real people." What is individually owned implies that others do not have rights of access to it except through the willingness or interest by the owner in sharing what is individually owned. The primary context for the sharing of what is individually owned is the family, however a family as a social unit may be culturally recognized and defined.

Individual ownership through individual procurement (such as vegetal foods obtained through foraging) is typically associated with resources that: (1) come in small units, hence may easily be shared within a family, (2) have a high probability of success for each foraging episode, hence are accessible by all able-bodied adults, and (3) may be obtained through skills possessed by all able-bodied adults. Most vegetal foods meet these criteria, and typically, vegetal foods foraged by women are individually owned and shared within the context of a family. Some animal resources also meet these criteria and may be individually owned after they are collected, snared or trapped.

In contrast to resources meeting these criteria are those that: (1) come in large units, and hence cannot easily be consumed by a single family, (2) have a lower probability of success for each foraging or hunting episode, and (3) the likelihood of success depends upon individual skills that are not equally shared by all able-bodied adults. Large animals generally meet these criteria, and an animal obtained through hunting of large animals typically is not owned by the hunter but is subject to cultural rules for sharing (Kelly 1995). These rules of sharing can range from relatively simple to highly formal. The Tiwi of Melville and Bathurst Islands of the northwest coast of Australia, the !Kung san of the Kalahari Desert in Africa, and the Netsilik Inuit along Hudson Bay in Canada illustrate the range of rules of sharing. For the Tiwi, sharing is determined by the position of the hunters in a boat when hunting for crocodiles. The !Kung san share the animal killed by a hunter according to cultural kin relations individuals have to the person owning the arrow that killed the animal (which can, but need not, be owned by the hunter) and based on obligations one has to one's kin such as a man providing bride service to his wife's parents. The Netsilik Inuit share seals obtained through the pack ice in the winter with sealing partners of the hunters as discussed above. Risk, taking into account both the likelihood and consequences of failure of a hunting episode, seems to determine the extent to which sharing is formalized and not left to the interests and decisions of the hunters. Among these three groups, the Tiwi have the least risk and the most informal system for sharing, whereas the Netsilik Inuit have the great-est risk and most negative consequences for failure, along with the most formal system of sharing. The !Kung san are in-between regard-ing risk and are more formalized than the Tiwi but less formalized than the Netsilik Inuit.

The !Kung san make it evident that a hunter is considered by the other members of the residence group—which consists of his close rela-tives—as acting as an agent for the group and not in his self-interest. Other members of the residence group go out of their way to downplay the accomplishment of a hunter. As one person expressed it: "When a young man kills much meat, he comes to think of himself as a chief or a big man, and he thinks of the rest of us as his servants or inferiors. We can't accept this. We refuse one who boasts for someday his pride will make him kill somebody. So we always speak of his meat as worthless. In this way we cool his heart and make him gentle" (Lee 1979: 246).

Hunting transforms what no one initially owns—that which is "in nature," such as animals whose roaming crosscuts the boundary of a !Kung san n!ore—into what is collectively owned and not into what is individually owned. The owner of the arrow is recognized as the agent of the collective group determined through kin relations and distrib-utes raw meat according to those kin relations and obligations, such as a man being obliged to provide meat for the parents of his wife. The recipients of the raw meat distribute it further as raw meat accord-ing to their kin relations and kin obligations. After this second dis-tribution, the meat is cooked, and cooked meat becomes individually owned, hence not subject to collective sharing. The cultural rules for meat distribution, then, transform what is corporately owned into what is individually owned, and this transformation is marked by raw meat being subject to cultural rules regarding the distribution of meat versus cooked meat being individually owned.[3] Like foraged food that is collectively owned "in nature" but becomes individually owned "in culture" by the action of the woman who has gathered the food, the action of women also transforms, through cooking of meat, what is collectively owned "in nature" into what is individually owned "in culture." In both cases, the actions of women transform what is in nature to what is in culture, whereas men's hunting acts on what is owned "in nature." Raw meat, the product of hunting, is still part of nature, implying the analogy, man:woman :: nature:culture.[4] The same analogy applies to the Netsilik modes of resource procurement. For the Netsilik, men are the procurer of resources and women are the managers of resources after they have been procured (that is, they transform what is "in nature" to what is "in culture"). The men hunt the seals as agents for the members of the winter sealing camp (hence they act on what is "in nature"), and the seal is butchered by the successful hunter's wife and distributed to the wives of his sealing partners according to the parts of the seal corresponding to their partnership: "it was the wives who performed the actual sharing" (Balikci 1970: 135)—hence transforming what is "in nature" to what is "in culture."

In all three cases, the sharing of hunted animals is not based on an individual decision by the hunter whether he should share, but takes place in a culturally specified manner. This takes decisions about sharing out of the hands, as it were, of the hunter(s) and thereby transforms what otherwise may be behavior that is unpredictable and/or subject to substantial variation, depending on individual interests in, or proclivity towards, sharing, into predictable outcomes when hunting has been successful. For the Netsilik Inuit, everyone knows in advance which families will get what part when a seal is killed; that is, they have culturally transformed what is individually at high risk into a system in which the risk is averaged out over all hunters, regardless of individual levels of skill, luck in hunting, and so on. Their system of rules for sharing also have the effect of reducing conflict that might arise over differences between what is expected and what is received. For the Netsilik, conflict could be fatal if it led to the breakup of a winter sealing camp. At the other extreme, the Tiwi limited sharing to those involved in the hunt, but the failure in one hunt did not pose high risk for the survival of an individual hunter and his family. Hunting could occur throughout the year in the tropical region inhabited by the Tiwi.

Egalitarian Hunter-Gatherer Societies

The Netsilik system of sealing partners exemplifies the general lack of hierarchy or social stratification in hunter-gatherer societies outside of a family or extended family context (Boehm 1999). Within a family or extended family, individuals are hierarchically related through older generation-younger generation relations. Kin relations to persons outside of an extended family are horizontal rather than vertical when conceptualized through living relatives rather than tracing to a common, deceased ancestor. Horizontal tracing occurs frequently with hunter-gatherer groups. For the Netsilik, the coherency of the *ilagiit nanagminariit* as a social unit was based internally on hierarchical, kinship relations among living kin, particularly father-son relations, and the social unit could fission when the eldest male who was its focal point died (Balikci 1970). Relations to persons outside of the *ilagiit nanagminariit* consisted of horizontal links formed through marriage and/or their system of sealing partners.

The system of sealing partners implied that the members of a winter camp collectively "owned" and had access to the parts of the seal killed by a hunter according to their sealing partner relationship to the hunter. Like the !Kung san hunter, an individual hunter acted as an agent for the group of sealing partners of which he was a member and not simply as an individual with his own set of interests. As a sealing partner, a Netsilik hunter could not assert individual ownership over the seals he killed and then parlay individual control and ownership of a kill into a hierarchical relationship vis-a-vis other group members through their need for meat. Opting out of the corporate group of sealing partners was not viable since a single hunter, or even the handful of closely linked male kin making up a social unit such as the *ilagiit nanagminariit,* would run the risk of starvation given the odds against obtaining seals through the pack ice by a single hunter. Consequently, access to resources "in nature" was collectively controlled through the cultural system of sealing partners making up a winter camp. The "payoff" to the individual hunter and his family was having access to seal meat through the average rate seals were killed by him and his sealing partners, thereby reducing drastically the risk to him and his family were they dependent solely on chance-driven variance in individual rates of killing seals.

Under these conditions, the absence of a hierarchical structure such as a dominance hierarchy or other form of social stratification that might be based on the skills or prowess of individuals is hardly surprising, regardless of whether humans are biologically inclined towards attempting to dominate other individuals when possible (Boehm 1999). There was an inherent power asymmetry between an individual male and the collective group of sealing partners that favored the collective interest expressed through the cultural system of sealing partners over individual interests. A culturally constructed social unit such as the set of sealing partners could control the behavior of an individual member who engaged in behavior they collectively saw as not being in the interests of the collective group; e.g., an attempt by an individual to establish a dominance relation vis-a-vis other group members. The group members had a shared interest in controlling the behavior of a group member attempting to act according to individual interests that were not in the interest of the group as a whole. Control could be through methods such as publicly expressed criticisms, gossip and social ostracization. Social ostracization could range from refusal to socially interact with the ostracized individual, to banishment from the group, or, in extreme cases, to execution (Boehm 1999). Within a social unit such as the Netsilik *ilagiit nanagminariit* composed of close cultural kin, social ostracization would be doubly powerful as it meant not only social isolation of an individual within the residence group, but isolation of that individual from his close kin around whom his life was centered.

Although the Netsilik are an extreme case since their survival depended upon adherence to a highly structured system for the distribution of seal meat as a way to deal with the risks introduced by the harshness of the conditions with which they had to cope, their system of sealing partners makes evident the tensions that may arise between the interests of individuals as individuals and the interest of individuals as members of a culturally constituted group. When we are a member of a culturally constituted group, we have a dual identity due, on the one hand, to the role and behaviors we take on as a group member versus, on the other hand, the identity we have as an individual expressed through our individual interests and behaviors. We see this duality in the way the risk associated with exploiting mobile resources is managed in hunter-gatherer societies at both the individual and the group level.

The degree of risk associated with hunting mobile food resources is variable across hunter-gatherer groups due to regionally different environmental and ecological conditions. Greater risk leads to increased variability in the quantity of resources procured per individual per unit of time. Risk management when exploiting large mobile resources comes into play through methods used to locate a mobile resource and, when located, through methods use to kill it, and when killed, through methods used to share the resource. Risk management occurs both at the level of individual behavior and at the level of the behavior of a culturally constituted group.

At the individual level, various means to reduce risk are employed such as cooperative hunting, sharing of information among hunters regarding animal movement, and the substantial time investment spent in becoming a skilled hunter (Kaplan et al. 2000). The last leads to differences in hunting skill levels, thereby causing some hunters to be consistently more successful than others when hunting. By itself, this opens the possibility that the greater access to large quantities of meat by the most skilled hunters could be parlayed into dominance relations, or other forms of social stratification, if individual interests are not countered by other group members acting in the interest of the collective group.

Hunting risk, measured by the likelihood of successfully killing an animal once it has been located, is reduced, among other ways, through increasing the complexity of the implements used in hunting as a way to ensure greater likelihood of killing an animal once it is located. For example, in comparison to simpler hunting weapons such as thrusting or throwing spears or spear throwers, the more complex bow and arrow is more versatile and reduces hunting risk (Hughes 1998). Not surprisingly, then, the complexity of hunting weapons is strongly correlated with degree of hunting risk (Torrence 1989; Read 2008b), with the latter varying inversely with the length of the growing season and ranging from relative low risk in tropical regions with a twelve-month growing season to high risk in Arctic regions with a short growing season. With longer growing seasons, there is a greater variety of animal species that can be hunted, making it more likely that at least one species will be encountered in a hunting episode. In addition, animals are accessible over a longer time duration, hence there is less risk than with a shorter growing season. Under low risk conditions, sufficient food resources can be obtained with simpler hunting weapons and there is little advantage to investing time in the making, maintenance and skill learning needed to use more complex hunting weapons. Under high risk conditions, controlling for risk depends on factors such as time spent on more frequent hunting episodes, increasing the chance that a hunt will be successful when a resource is met, and so on. Hunting success is made more likely through use of more complex hunting weapons that increase the likelihood of a mobile resource being captured or killed after it is located.

Because increased risk, coupled with difference in skill levels, leads to increased variability across time and among hunters in the amounts obtained by a hunter, cultural systems of resource sharing among the members of a group have the effect of distributing risk across group members by evening out variation in the quantities of resources obtained by a single hunter. Cultural rules for sharing have the effect of removing sharing from individual decision making and making it part of one's identity as a member of a culturally constituted group. A Netsilik hunter does not decide whether it is in his interest to share a seal he has killed, but shares it in the prescribed manner since to do otherwise contradicts his understanding of what it means to be a sealing partner. The expectation of his sealing partners that they will receive portions of the seal through their respective wives and according to their relation to him as sealing partners reinforces his sharing according to his role as a sealing partner. To act otherwise is equivalent to him saying that he (and his family) are opting out of the system of sealing partners.

Opting out is equated with becoming a free-rider in genetic theories that attempt to identify sharing behavior as the phenotypic expression of an underlying genotype determined by a so-called "cooperate" gene and where free-riding is taken as the phenotypic expression of an underlying genotype determined by a so-called "free-riding" gene. From a genetic viewpoint, individuals with the "free-rider" phenotype benefit from the behavior of those with the "cooperate" phenotype but not vice-versa, hence the "free-rider" gene will win out against the "cooperate" gene under natural selection. However, the consequences of opting out of the system of sealing partners is not captured by saying the hunter has become a free-rider, for that assumes he will still be the recipient of seal meat obtained by other hunters. Instead, he has not only redefined his relation to the other hunters, he has also redefined the relation of the other hunters to him. From their perspective, he has rejected, as it were, the cultural system of which they are a part. Their cultural system defines for them appropriate, mutually understood behaviors that are part of what it means to be a sealing partner.

The system of sealing partners has the superficial appearance of what is called reciprocal altruism (Trivers 1971). The sharing behavior is considered to be altruistic since each successful hunter gives up seal meat that otherwise would be used as food for himself and his family. It is reciprocal since the receiving hunter also gives seal meat to the donor hunter when he kills a seal. However, the behavior is not the impetus for, but the consequence of, the system of sealing partners. Opting out of the system of sealing partners will likely invoke a negative response on the part of a sealing partner not so much because it interrupts the system of reciprocal altruism and thereby calls for a tit-for-tat response such as "if you do not share with me I won't share with you," but because it violates their cultural system of rules that they hold in common for the behavior of sealing partners. Failure to act properly violates their system of cultural rules that is part of their social identity. This can lead to social disruption and a violation of the understood social order. The result may be collective punishment of the violating individual as a way to restore the social order. For example, among the !Kung san, "groups punish to exert sanctions against those who break norms ... [and when] enforcement of the obligations of kinship and joint landownership that bind individuals into groups was not sufficient to maintain community in the face of disruptive forces ... the goal of the Ju/'hoan was to bring the transgressors back in line through punishing, without losing familiar and valuable group members" (Wiessner 2005: 134, 139).

It should be noted, though, that just as the hunter who refuses to share a seal he has killed is making a choice between acting according to the culturally constructed role of what it means to be a sealing partner or acting in his self interest, a sealing partner has the same choice in response to that hunter's action. A sealing partner can respond either in his role as a sealing partner to the hunter acting in violation of their mutual understanding of what it means to be a sealing partner, or in accordance with his self-interest; e.g., he has not received a portion of the seal killed by the hunter and so he might make a tit-for-tat response and refuse to share seal meat when he has killed a seal, thereby further contributing to the breakdown of the social order. For the winter sealing camps, the breakdown of the social order would likely lead to death through starvation, hence all the families in the camp have a vested interest in maintaining the social order and in bringing sanctions against those who act in violation of the social order they have defined for themselves.

Summary

Kinship terminologies are prototypical examples of "culture as a complex whole." The terminology expresses the way a group of individuals are structured vis-a-vis each other according to mutually recognized kin relations. This kind of structure defined through the kinship terminology is central in the odyssey from non-human primate societies to human societies.

A kinship terminology is *cultural* because of being constructed and by being a construction with properties and the meaning of kinship relations known to all societal members. It is *complex* because of being a system that enables users of the terminology to make calculations of kin relations through symbolic representation of kin relations. It is a *whole* because it is a set of kin terms interconnected internally as a system and bounded conceptually through the computation of other kin terms through kin term products. We cannot dissect a kinship terminology into its individual terms and then consider each term as a separate trait to be transmitted individually, with evolutionary change characterized by change in the frequency distribution of each trait. Instead, the terms are labels for positions in a system of kin relations, and what is transmitted is that system of relations. The labels for the positions—the kin terms themselves—can change, or we can have variant labels such as mom, dad, papa, mommy, my old man, etc., for a position, but what is constant is a kinship terminology composed of a structured system of positions. As members of the same cultural milieu, we agree on what the positions are and we agree on a standard labeling for those positions. Over time, parts of the kinship terminology structure can change—evolution can take place; it is not evolution measured by change in the frequency of individual traits making up this structure, but rather by change in the organizational structure of the system of kin terms.

Notes

1 Another example is the genetic trait known as sickle-cell anemia in which red blood cells take on a sickle-shape and become less efficient in transporting oxygen to the body cells. The co-evolution was due to slash and burn farming made possible and introduced into tropical Africa by the use of iron plows. Slash and burn agriculture established conditions under which malaria could become endemic (Livingston 1958). The trait was selected for as it provides protection against obtaining malaria, thereby compensating for the negative effects of the anemia caused by the sickle-shaped blood cells.

2 A distinction is sometimes made between Immediate Return (IR) and Delayed Return (DR) hunter-gatherer societies (Woodburn 1982; Stiles 2001), where IR societies are those that consume food resource upon procurement and DR societies are those that engage in some form of food storage. Generally, DR societies are the more complex societies we are excluding here, although some of the small-scale, arctic hunter-gatherer societies fit the DR definition due to highly seasonal resources that required food storage for times of the year when food was otherwise not available. Though these are DR societies by definition, food storage in the small scale, arctic hunter-gatherer groups is primarily a means to survive in a region without year-round food sources rather than being the basis for maintaining more complex forms of social organization.

3 The raw/cooked opposition for meat among the !Kung san exemplifies the opposition discussed by Lévi-Strauss (1973) in his book *Le Cru et Le Cuit* (The Raw and the Cooked).

4 This runs contrary to the argument made by Sherry Ortner (1974, 1996) that universally man:woman :: culture:nature.

CHAPTER 4

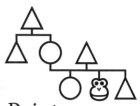

The Chimpanzee Middle Point

Temporally and phylogenetically the chimpanzees (*Pan troglodytes* and *Pan paniscus*) are between the OW monkeys and *Homo sapiens* hunter-gatherers. There is general agreement that an ancestral species similar to the ancestral species of the OW monkeys gave rise, evolutionarily speaking, to an ancestral species for both the species in the genus *Pan* and eventually to our species, *Homo sapiens*. The evolutionary development of modern *Homo sapiens* involved adaptive changes that took place in eastern Africa, the Middle East, and the European subcontinent over the past several hundred thousand years. These changes eventually led to anatomically modern *Homo sapiens* fully in place by 50,000 years ago, with forms of social organization comparable to those of recent hunter-gatherers.

The extensive overlap in the DNA of *Pan troglodytes* with *Homo sapiens* suggests continuity in trait change for this evolutionary trajectory. Continuity in morphological traits is also implied both by evidence regarding the evolution of quantitative traits and by fossil evidence. The development of bipedalism, the hallmark for our anatomical distinctiveness among the primates regarding locomotion, led to quantitative differences in the metric dimensions of the parts of the skeletal structure affected by this change in locomotor pattern. This quantitative change did not lead to a qualitative disjunction in the anatomical differences between modern *Homo sapiens* and a species ancestral to ourselves and modern day chimpanzees. Similarly, a comparison of the brain structure between ourselves and the chimpanzees shows primarily quantitative, not qualitative, differences. This suggests, to the extent behaviors can be related to the operation and functionality of the neurological structures of the brain, that we should find continuity in behaviors and not qualitative disjunction. Finally, the fact that we account for evolution from a common ancestor to the chimpanzees to *Homo sapiens* through the same, incremental biological evolutionary process we invoke for evolutionary change in other organisms suggests that the changes are at a quantitative, not a qualitative, level.

Nonetheless, biological evolution can also lead to qualitative differences among organisms such as the change from single cell to multi-cellular organisms or the change from non-sexual to sexual forms of reproduction. Less dramatically, we can consider the shift from a gill-based to a lung-based system of oxygen intake or the shift from a primarily terrestrial to an arboreal mode of locomotion or the shift from a nocturnal to a diurnal mode of adaptation by the primates to be qualitative changes. In each of these examples, the change also led to new forms of species based on modes of adaptation that were not simply an elaboration on prior modes of adaptation. The question I pose in this chapter, then, is whether, in the transition from the OW monkey mode of adaptation to the chimpanzee mode of adaption, we see a pattern of change that, when extended, leads to *Homo sapiens* and a hunter-gatherer mode of adaptation? Or do we find discontinuity in the evolutionary trend going from an OW monkey mode of adaptation to a chimpanzee mode of adaptation and then to the mode of adaptation that characterizes hunter-gatherers? The argument I make in this and the next chapter is that we find discontinuity characterized by a transition from an experiential to a relational basis for social organization and social coherence as we go from the non-human primates to modern *Homo sapiens*. The discontinuity arises through the evolutionary appearance and development of constructed systems of kinship relations discussed in the previous chapter and upon which the kinship relations among individuals and the coherency of social groups are constructed. These relations are not determined by prior interactions among individuals but by a conceptual system that determines connections among societal members—whether physically present or not—upon which social interaction both depends and is enabled.

To show this, we first consider the changes in the mode of social systems that took place with the evolutionary origin of the chimpanzees. Then we examine whether the trends I identify, if extended, would lead to a hunter-gatherer social systems similar to the ones I discussed in the previous chapter. In making this argument, I will focus primarily on social organization, for it is in the social domain where we would find discontinuity, if at all.

Performative Versus Ostensive Forms of Social Organization

The trend we identify in going from the OW monkeys to the chimpanzees centers on a dimension for characterizing forms of social organization discussed by Shirley Strum and Bruno Latour (1987) as a way to account for variation in the social organization of baboon societies. The dimension they identify goes from *ostensive* to *performative* forms of social organization. At one extreme of this dimension, ostensive forms of social organization are those for which it "is, *in principle*, possible to discover the typical properties of what holds a society together," and accordingly social "actors ... are *in* the society ... [and] their activity is restricted because they are only part of a larger society" (785, emphasis in the original). At the other extreme, performance societies are those where "*in practice* actors ... define, for themselves and for others, what society is, both its whole and its parts" (785, emphasis in the original). In an ostensive society, individuals are born and enter into an already existing and established system of social organization. The form and content of social interaction among individuals is constrained by, but not determinative of, that social organization. A performative society, however, is one in which the form of social organization structure emerges out of the interactions among the societal members and hence can vary in substantive ways from one context to another according to differences in how patterns of interaction vary from one context to another.

Societies with behaviors determined genotypically will correlate with the ostensive end of this dimension. Ostensive societies would include OW monkey species with female philopatry and social systems based on matrilines internally structured through females with linear dominance relations. The OW monkey species also have performative aspects to their social organization since the linear dominance hierarchy is worked out according to interactions involving biological mother-biological daughter relations. Under these conditions, biological kin selection, sexual selection, and female parenting selection are all active processes, each of which leads to behaviors having phenotypic expression correlated with an individual's genotype. Similarity in genotype characterizes biologically kin related individuals, and this will lead to commonality in behaviors. For example, biological mothers in OW monkey species with a matrilineal and dominance hierarchy form of social organization appear to be predisposed genetically to support their biological daughters in agonistic encounters with other females in the troop. This leads, as discussed in chapter 2, to the perpetuation and stability of a form of social organization centered on a female dominance hierarchy.

Performative primate societies, in contrast, will have greater intra-species variability in social organization, and social behaviors will be less affected by selection processes such as biological kin selection. To put it another way, individualization of socially relevant behavior will be more prevalent in performative societies, or perhaps more accurately, selection for individualization of social behaviors will push the social system in the direction of a performative social system so that the form of social organization is less determined across different social units, thereby leading to greater variability in intra-species forms of social organization.

Variation in Social Organization among the Lesser and Greater Apes

Perhaps one of the most striking differences we see in the transition from OW monkeys to the apes lies in the high diversity of forms of social organization we find among the greater and lesser apes. As we saw in chapter 2, the relatively limited number of forms of social organization among the OW monkeys are repeated across different species. In particular, a form of social organization that we can define ostensively as characterized by the presence of female philopatry, matrilineal units and a stable, linear dominance hierarchy is repeated across a wide variety of species and genera of OW monkeys. OW monkey species with this form of social organization are found in virtually every climatic and environmental zone of planet earth. Social organization like this appears to be highly effective in that it can accommodate itself to widely different environmental circumstances.

In contrast, each of the genera, and even species, making up the lesser and greater apes has a distinct form of social organization. In the southeastern Asian area, the gibbons (*Hylobatid*) and siamangs have pair bonding and a monogamous form of social organization. Orangutans (*Pongo*) have a solitary form of social organization with adults living in the same area over long periods of time (Harcourt and Stewart 2007, and references therein). In Africa, gorillas (*Gorilla*) are characterized by stable social units typically made up of a single male with a group of females and offspring, or *harem* form of social organization (Harcourt and Stewart 2007, and references therein). In the *Pan* genus, *Pan troglodytes* has multi-male, multi-female communities with male philopatry. Males in these communities are characterized by a fission-fusion form of organization (with frequent change, even within hours, in the composition of male social units), whereas females have a less gregarious form of organization than males (Harcourt and Stewart

2007, and references therein). *Pan paniscus* is also male-philopatric, but social units are organized around females rather than males and there are strong biological mother-biological son ties (Harcourt and Stewart 2007, and references therein). In short, there is no typical form of social organization for the apes.

Variation in Chimpanzee Social Organization at the Species Level

In contrast to the OW monkeys, the wide variation in forms of social organization across the genera making up the lesser and greater apes suggests that no single model adequately provides a baseline pattern for their varied forms of social organization. Even within the *Pan* genus, the model for a troop structure for the OW monkeys (see Figure 2.2) does not account for the forms of social organization occurring in the two species, *Pan troglodytes* (Pt—common chimpanzee) and *Pan paniscus* (Pp—"pygmy" chimpanzee), in this genus. Though both are male philopatric, each lacks a troop form of social organization and has no coherent social unit encompassing males, females and offspring on a day-to-day basis as would be implied from a model for the social organization of the OW monkeys. Instead, each of the *Pan* species lives in communities internally organized on a fission-fusion basis forming temporary social units (Nishida 1979), with marked differences in social organization between the two species. Pt is generally characterized as male gregarious, male bonded, and with a social structure centered on male relationships based on grooming, whereas the Pp social structure is centered on female relationships based on sexual behavior in the form of genital rubbing. While both are territorial, they also differ strikingly in that Pp communities defend a territory without lethal violence whereas Pt male coalitions aggressively use lethal violence both to defend their territory (Wilson and Wrangham 2003) and to expand its size (Mitani, Watts, Ambersi 2010). The ability to use lethal violence effectively as part of territorial defense seems to relate to the behavioral and cognitive capacity of the Pt chimpanzees to form male coalitions that can attack an isolated male in a neighboring community with little risk to the attackers (Wilson and Wrangham 2003).

Variation in Chimpanzee Social Organization and Behavior at the Community Level

Even within a single *Pan* species there is considerable variation in social organization across communities. The fission-fusion social units, for example, have different sexual compositions across the Pt communities. In the Gombe Stream communities in Tanzania, males form small, unstable social units, and the females are largely solitary (Wrangham and Smuts 1980). In the Taï Forest communities in the Ivory Coast, however, the social units are often composed of males and females (Boesch 1966). Further, the asociality of females found at Gombe and other East African chimpanzee (*Pan troglodytes schweinfurthii*) communities (Wakefield 2008) differs in degree. The females in the West African chimpanzee (*Pan troglodytes verus*) communities are more social than those in the East African communities (Lehmann and Boesch 2008). In addition, the female pattern for social interaction in the East African communities is more complex than their asociality suggests. The females in the Ngogo, Uganda, community in East Africa, for example, while having a low overall average rate (7 percent) of female interaction, which is consistent with other East African communities, form non-interacting cliques within which the rate of female gregariousness (17 percent) is comparable to the rates found among Pp females (25 percent) and West African Pt females (19 percent) (Wakefield 2008: Table III, and references therein).

Male and female chimpanzees also differ through the home range of females being smaller than the home range of males, which is in accord with males defending a feeding territory exploited by both males and females. The more limited home range for females relates (along with females having smaller foraging group sizes than males) to resource competition among females (Pokempner 2009, and references therein), although the spatial relationship of the female range to that of the males (a sub-home range or an independent range) is still unclear (Wilson and Wrangham 2003).

Variation in Grooming Behavior

The chimpanzees also differ markedly from the OW monkeys regarding grooming patterns. For the Pt chimpanzees, grooming centers around adult male-adult male grooming (Nishida 1979) and is multidimensional in comparison to grooming in the OW monkeys, including 'cultural diversity' in how grooming is done across communities: (1) handclasp grooming (practiced by chimpanzees in the Mahale Mountain National Park, Tanzania) where the grooming individuals also clasp and hold their free hand above their heads (McGrew and Turin 1978); (2) social scratch (practiced by the Mahale chimpanzees and in Ngogo, Uganda) where the grooming chimpanzees simultaneously scratch each other

(Nakamura, McGrew and Marchant 2000; Nishida, Mitani and Watts 2003); and (3) 'lip smacking' while grooming (practiced by chimpanzees in Mahale and Gombe chimpanzee communities) (Nishida, Mitani and Watts 2003). In addition, chimpanzees often groom face-to-face, unlike other primates (Nakamura 2003). In Gombe, both grooming and higher rates of aggression occur between males when there is fusion of male groups (Bauer 1979; Aureli and Schaffner 2007, and references therein), whereas a lower rate of forming social units through fission-fusion among PP males correlates with a lower rate of mutual grooming (Furuichi 2009) and greeting behaviors (Furuichi and Ihobe 1994).

In contrast with the dyadic and unidirectional grooming of OW monkeys, polyadic grooming (in which more than two individuals groom each other simultaneously) is common among chimpanzees, and grooming is often bi-directional, hence reciprocal (Nakamura 2000). Bi-directional grooming appears to relate to forming bonds between males, but it is not a form of reciprocal altruism (Fedurek and Dunbar 2009), and so bi-directional grooming does not appear to be the outcome of biological kin selection. Instead it is a performative, rather than an ostensive, behavior. This is consistent with experimental evidence finding weak or no evidence for chimpanzees engaging in contingent reciprocity wherein the subject preferentially interacts with a partner who has recently helped the subject (Melis, Hare, and Tomasello 2008; Brosnan et al. 2009; see also Cheney et al. 2010 for similar experimental outcomes on female baboons).

Like OW monkey species in which troop members reciprocally exchange grooming and provide coalitionary support, chimpanzees at Ngogo also make reciprocal grooming exchanges and provide coalitionary support (Schino and Aureli 2009, and references therein). In both cases, reciprocity uses the same currency: grooming for grooming and coalitionary support for coalitionary support. In addition, the Ngogo chimpanzees exchange meat obtained from hunting. Unlike other primates, though, the chimpanzees at Ngogo also make reciprocal exchanges using different currencies. They exchange grooming for coalitionary support and, non-reciprocally, coalitionary support for meat sharing (Mitani 2006, and references therein).

Male Social Units and Dominance Hierarchy

At Ngogo, most (69 percent) of the coalitions are composed of males and the coalition partners each outrank an opponent in the dominance hierarchy (Watts 2002). Male coalitions are fluid and a male may turn against his former coalition partner (Muller 2002). Despite the fluidity, coalitions are important for challenging dominance relations since "having a coalition partner may raise a male's confidence to the point that he challenges a higher-ranking individual" (Muller 2002: 121). Coalitions like this are destabilizing for a dominance hierarchy and seldom occur in OW monkeys (Boesch and Boesch-Achermann 2000).

Occupying the alpha male position in a male dominance hierarchy can be stable over several years; nonetheless, high ranking males are aggressive and act as if their position is potentially unstable: "a high-ranking male can never be certain what political maneuvering has occurred in his absence [due to changes in social units], it is necessary for him continually to reestablish his dominance when parties come together.... The large proportion of aggression ... takes place in the context of reunions" (Muller 2002: 121). From the perspective of the ostensive/performative dimension, these grooming behaviors have performative consequences, since a coalition structure emerges from interactions such as mutual grooming between males.

At the community level, the mean size of social units varies from 2.4 to 8.6 adult individuals across ten Pt communities (Lehmann, Korstjens, and Dunbar 2007). Two other communities, one at Mahale (Furuichi 2009: Table 1) and the other at Okoro Biko (Jones and Sabater Pi 1971), have larger social units with means of 10.3 (Furuichi 2009: Table 1) and 11.2 (Jones and Sabater Pi 1971) individuals, respectively. In contrast, the Pp social units tend to be larger than the Pt social units, with means ranging from 6.7 to 17.9 for three communities (Lehmann, Korstjens, and Dunbar 2007; Mulavwa et al. 2008, and references therein). This difference in size of social units between Pp and Pt and within Pt does not, however, correlate with resource availability as happens with variation in foraging group size among the OW monkeys. Nor does the fruit-patch size difference between the regions exploited by Pt (Kibale, Uganda in East Africa) and Pp (Lomako, Democratic Republic of the Congo) account for differences in their respective social unit sizes (Chapman, White, and Wrangham 1994).

Relationship between Social Unit Size and Food Resources

As for the differences in social unit size within the Pp communities, the data relating social unit size to resource abundance are equivocal. White (1996) shows a positive correlation between social unit size and food patch size for the Pp communities in Lomako, but Hohmann and

Fruth (2002) report that social unit size does not vary with amount of available ripe fruit. The relationship varies positively only for preferred fruits in the communities in Kahuzi-Biega, D. R. Congo (Basabose 2004), and correlates positively, but with little change in total group size, for communities in Wamba, D. R. Congo (Mulavwa et al. 2008). There appears, then, to be only a limited relationship between social unit size within the Pp communities and food abundance.

Data on fruit abundance and social unit size for Pt communities are equally varied, with some studies showing that larger social units are formed in months with higher fruit abundance, whereas other studies find the contrary (Furuichi 2009, and references therein). Similar to the Lomako Pp communities, there is a positive relationship between social unit size and patch size but not between social unit size and food abundance for the Sonso Pt community in Budongo Forest Reserve, Uganda (Newton-Fisher, Reynolds, and Plumptre 2000). Social factors appear to have a confounding effect as the presence and number of estrous (sexually receptive) females correlates with fruit abundance (Newton-Fisher, Reynolds, and Plumptre 2000; Hashimoto, Furuichi, and Tashiro 2001). This, coupled with the observation that the size of social units is larger when estrous females are present (Matsumoto-Oda et al. 1998; Mitani, Watts, and Lwanga 2002), implies that presence or absence of estrous females may be a confounding factor when correlating unit size with fruit abundance, and the failure to take this into consideration could account for some of the differences reported in the relationship between fruit abundance and Pt unit sizes (Hashimoto, Furuichi, and Tashiro 2001, and references therein).

Overall, these data show lack of a clear relationship between fruit abundance and unit size. This, along with the fact that social units change in the Pt communities on a time scale of hours or even minutes (Boesch and Boesch-Achermann 2000: Table 5.1), suggests that social units are less a part of an ostensively definable social structure responding to external factors and more the consequence of social structure determined through performance on a day-to-day and hour-to-hour basis and responding to internal factors.

Performative Basis of Male Social Units

This conclusion regarding social structure determined through performance is supported by evidence showing that male chimpanzees do not bias social behavior such as association, grooming, proximity, coalition, meat sharing, and patrolling towards biological kin (biological siblings, biological cousins, etc.) as might be expected under biological kin selection (Mitani 2006a). Instead, the only bias is towards maternal, but not paternal, biological siblings, despite male philopatry. In addition, individuals who engage in cooperative actions more often than would be expected by chance are substantially more likely not to be kin-related (Langergraber, Mitani, and Vigilant 2007). These authors attribute the pattern to "individuals maximizing their own fitness by cooperating with age mates" (2007: 7788). For both males and females, biological "kinship plays a limited role in structuring the intrasexual relationships" (Langergraber, Mitani and Vigilant 2009: 840) despite the presence of long-term social bonds between maternal biological siblings and half-siblings (Mitani 2009). This may also be seen through the fact that while social organization is typically based on dyads, in at least one community (Ngogo) the males appear to be organized at a higher level than dyads into two subgroups—stable for the three years of observation—formed according to age and dominance ranking and not maternal biological kinship (Mitani and Amsler 2003). Although the long-term social bonds structured around maternal biological kin such as maternal siblings are analogous to similar patterns found in OW monkeys, chimpanzee social bonds are also established through interactions with age mates (who need not be biological kin), which implies a performative basis for forming social bonds.

Within this framework of varied social interactions, grooming by chimpanzees in Budongo Forest, Uganda, appears to take place based on "the current and future state of relationships between males" (Newton-Fisher 2002: 134); that is, as a way to create social structure through interaction rather than as an expression of already existing social structure. Their interactions are directed towards a "struggle for high social status within a community" (Newton-Fisher 2002: 125). Males do this by forming coalitions involving two or more individuals involved in aggressive actions against another individual, as well as coalitions involved in inter-community agonistic encounters (Mitani, Merriwether, and Zhang 2000, and references therein). This contrasts, for example, with the simpler pattern for coalitions in OW monkeys in which a dominant female intervenes in an agonistic encounter between two females on the side of the more dominant female of the two females.

From Old World Monkeys to Chimpanzees

Table 1 summarizes the properties and behaviors relating to social organization and social structure discussed above that have changed or been modified in the transition from OW monkeys to *Pan*. In all instances, we see elaboration, expansion and/or innovation in comparison to the properties and behaviors that characterize the OW monkeys, along with a reduction in the phylogenetic scale on which we find substantive differences. Whereas properties and behaviors are found in common across various genera of the OW monkeys—referred to as "marked uniformity in patterns of social organization revealed among the Old World monkeys" (Di Fiore and Rendall 1994: 9943)—properties and behaviors have significant inter- and intra-species variability at the species and community level in the genus *Pan* (Doran et al. 2002). Although intra-species variation does occur in OW monkeys such as baboons (Strum and Latour 1987, and references therein), it does not occur at the same scale or to the same extent as in Pt or Pp. Behaviors in one *Pan* community differ markedly from behaviors in another community, and even within the same community patterns of social interaction can vary significantly from one period of observation to another (Boesch and Boesch-Achermann 2000).

Old World Monkeys and Ostensive Patterns of Social Organization

The "uniformity in patterns of social organization" mentioned by Di Fiore and Rendall underscores the way the OW monkeys' social organization fits on the ostensive side of the ostensive-performance dimension. This is not to say that performative behaviors leading to an emergent structure are absent in the OW monkeys, but rather that performative behavior takes place in the OW monkeys within a framework bounding the scope of such behaviors. A newborn monkey female (in a female philopatric species) will eventually be slotted into the female dominance hierarchy below her biological mother and above her older biological sisters because of support provided to her by her biological mother (and possibly other females in her matrilineal unit) in agonistic encounters involving lower ranking females (Chapais 1992). Her position in the dominance hierarchy emerges from these social behaviors undertaken by her biological mother and other maternal kin, hence from performative behavior. At the same time, the fact that a biological mother will intervene on behalf of a biological daughter in agonistic encounters is part of a background framework already in place and does not emerge from performative behavior. As a consequence, the stability of the dominance hierarchy does not emerge from performance alone, otherwise dominance hierarchies among male chimpanzees would be equally stable, but they are not. Since the same, general form of social organization

Table 1: Comparison of Old World Monkeys and Chimpanzees

Old World Monkeys	Chimpanzees
1. Troop a. Coherent group b. Stable social units based on matrilines	1. Community a. Fission-fusion group b. Unstable social units over short time frames c. Social organization of males distinct from that of females d. Sexual composition of social units differs among communities
2. Grooming a. Unidirectional grooming b. Dyadic grooming c. One dimensional grooming d. Grooming focuses directly on biological kin via matrilineal unit	2. Grooming a. Bi-directional (reciprocal) grooming b. Polyadic grooming c. Multidimensional grooming d. Grooming focuses indirectly on biological kin via age mates e. "Cultural" variation between communities in grooming style
3. Coalitions a. Provide access to sexual partners b. Occur between biological kin	3. Coalitions a. Provide access to sexual partners b. Occur between biological kin and non-biological kin c. May be destabilizing
4. Stable linear dominance hierarchy a. Dominance position established through mother-daughter support b. Low rates of aggression, act in accordance with position in dominance hierarchy	4. Contestable linear dominance hierarchy a. Coalitions used to establish position in hierarchy b. High rates of aggression by high ranking males
5. Group size related to ecological constraints	5. Social unit size weakly related to ecological constraints
6. Biological kin selection a. Stable social units of biological kin provide conditions for biological kin selection b. Social behaviors are biased towards biological kin	6. Biological kin selection a. Lack of stable social units makes biological kin selection less likely b. Many social behaviors are directed toward age mates c. Cooperative behavior directed towards maternal siblings but not paternal siblings

Chapter 4

repeats itself across OW monkey species from different genera and is adapted to widely different ecological conditions, it transcends the performative interactions among troop members, hence the social organization of the OW monkeys lies on the ostensive side of the ostensive/formal structure to performance/emergent structure dimension.

Chimpanzees and Performative Patterns of Social Organization

With *Pan*, rather than the coherency of an ostensibly definable and integrated social system that subsumes individual behaviors (Strum and Latour 1987), we see instead social relations in which individual chimpanzees "define, for themselves and for others, what society is, both its whole and its parts" (Strum and Latour 1987: 785). We can see this in the fission-fusion pattern for the community that, contrary to a troop composed of matrilineal units in which individuals are embedded, is built up of unstable social units. Each of the social units is the outcome of the actions of individual chimpanzees and lasts only as long as the individuals in a unit continue to associate with one another, in some cases "changing composition every twenty minutes" (Boesch and Boesch-Achermann 2000: 261). For the community as a whole, territorial defense is through lethal violence against a neighboring community and, rather than being the collective action of the community, emerges from intensive social interaction by some of the males that leads to a coalition of males acting with singleness of purpose as a border patrol and killing individuals they encounter from a neighboring community.

Within the community, a male linear dominance hierarchy emerges from male interaction (Watts 2002, and references therein), but there is no overall, structural pattern regarding who becomes dominant over whom as with the OW monkeys. Because the dominance hierarchy emerges from performance, it is subject to change through performance, including challenges of an alpha male by a coalition of lower ranking males. As a consequence "dominance relationships and power relations must be evaluated anew after each social change" (Boesch and Boesch-Achermann: 2000: 265). The coalitions, then, are the emergent outcome of bonding among male chimpanzees arising from grooming dyads that crosscut biological kin and non-biological kin.

Social Brain Hypothesis and Social Complexity

Coalition formation with non-biological kin as well as biological kin leads to individuals interacting intensively with a lager cohort than when interaction is focused primarily on biological kin. According to

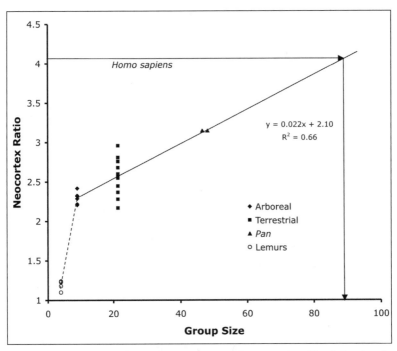

Figure 4.1 Change in the neocortex ratio with group size. The left triangle corresponds to *Pan paniscus* (mean group size based on mean community sizes in *n* = 2 study sites), and the right triangle to *Pan troglodytes* (mean group size based on mean community sizes in *n* = 11 study sites). Solid line: regression of neocortex ratio on group size for arboreal and terrestrial OW monkeys and *Pan* ($p < 0.0001$). The group size of *n* = 92 predicted for *Homo sapiens* is based on a neocortex ratio of 4.12 for early modern *Homo sapiens*. Data on community sizes are from Lehmann, Korstjens, and Dunbar 2007b. Data on neocortex ratios for Homo sapiens are from Aiello and Dunbar 1993. Data on neocortex ratio and group size are from Kudo and Dunbar 2001: Table 1.

the social brain hypothesis that attributes brain expansion to increasingly complex social relations (Dunbar 1992), the neocortex ratio for *Pan* should be greater than the ratio for the OW monkeys. Data on the neocortex ratio corroborate this expectation. The neocortex ratio for each of the *Pan* species is greater than the neocortex ratio for any of the OW monkeys (see Figures 2.8 and 4.1).

Neocortex Ratio Versus Intensity of Interactions

The complexity of the social group can relate either to group size through the size of a cohort of interacting individuals increasing with the group size, or to increased intensity and diversity of interactions

Chapter 4

within the same-size group, or to a combination of these. The increase in social complexity indexed by the differences in the neocortex ratios shown in Figure 2.8 relates to the intensity and diversity of interactions and not to the troop or community size per se for the following reasons. First, the troop sizes are only slightly larger when going from arboreal to terrestrial OW monkeys, and community sizes are comparable to the upper end of troop sizes when going from terrestrial OW monkeys to chimpanzees (Pp or Pt). In addition, social complexity tends to be relatively invariant regarding troop size for the OW monkeys due to social organization based on matrilineal units organized as a stable, linear dominance hierarchy. As a result, variation in group size relates to factors, such as density and distribution of food resources, that have little direct effect on social complexity. This implies that we can use the smallest troop size in each of the two OW monkey groupings as a proxy for social complexity for that grouping. For the chimpanzees, though, social complexity relates to the frequent formation and dissolution of social units and consequent interaction between and among individuals in social units (Newton-Fisher 1997; Aureli 2008), although the extent to which individuals associate with each other is independent of community size (Lehmann and Boesch 2004). Consequently, the larger communities are more complex than the smaller communities, and we therefore should use the mean community size as a proxy for social complexity. We will keep the mean community size for each of Pt and Pp distinct because of the differences in the social organization between the two species.

Predicted Group Size for Upper Paleolithic

A graph of the neocortex ratio versus group size shows a linear relationship with change in the neocortex ratio (see Figure 4.1). The linear trend, when extended, implies a group size of about 90 persons for the level of social complexity corresponding to the neocortex ratio = 4.12 for the Cro-Magnon, early modern *Homo sapiens* fossil dating to 30,000 BP (Aiello and Dunbar 1993, and references therein).[1] We can test this prediction by converting the settlement area of actual sites to an estimated range of settlement areas corresponding to a population size of $n = 90$. Good published data on site areas are available for the Solutrean sites. Since they are later sites and date to 22,000 to 17,000 BP, we need to allow for the possibility that during this later period maximal group size may been slightly larger than at the earlier date of 30,000 BP.

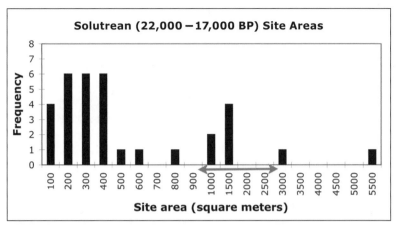

Figure 4.2 Histogram of *n* = 33 Solutrean sites from the Périgord region in France. The estimated range of site areas corresponding to a population size of 90 persons is indicated by the double-headed gray arrow. *Data on site areas are from White 1982.*

We make the conversion from site area to population size by using a conversion factor of 10 - 30 m^2 per person (Hassan 1978). This yields an estimated band of site areas varying from 900 to 2700 m^2 for *n* = 90 persons. This band of site areas corresponds to the middle portion of the overall range of Solutrean site areas (see double headed arrow in Figure 4.2). The middle portion corresponds to a second tier of Solutrean site areas that may be "aggregation sites for otherwise dispersed local groups" (White 1982: 88), thus sites with more complex social relations than occurs with the division of the population as a whole into the smaller, dispersed social units. Aggregation sites like this are absent from the European Middle Paleolithic (Mellars 1976, 1982; White 1982). All together, and even allowing for slightly larger group sizes, the site size data for the Solutrean sites support the predicted value of around 90 persons for the maximal size of groups.

Social Complexity and Individualistic Behavior

At first glance, this trend towards increased cognitive capacity, measured by the neocortex ratio, as an adaptive response by individuals coping with increased social complexity seems to suggest continuity in the evolutionary pathway from non-human primates to *Homo sapiens*. The matter, though, is more complex as we need to consider the basis for the increased social complexity. In a proximate sense, we can relate the increased complexity we find in *Pan* in comparison to the OW monkeys to the specifics of their respective forms of social organization and social structure. Here we see a trend towards a more complex social world through greater diversification of behavior: "Behavioural diversity is high in wild chimpanzees, and no population can be considered as the prototype for the whole species" (Boesch and Boesch-Achermann 2000: 260). However, biological kin selection as an ultimate cause has not expanded in scope among the chimpanzees beyond what occurs with the OW monkeys.

For the chimpanzees, cooperative behaviors that might arise through biological kin selection are biased only towards maternal biological siblings and not towards paternal biological siblings or more distantly related maternal biological kin, despite male philopatry that keeps paternal biological kin in the same community. Absent expansion of the biological kin field involved in biological kin selection, expansion of the scope of social behaviors to a wider cohort depends increasingly on intensive face-to-face interaction structured by factors such as dominance, age relations and familiarity rather than biological kin relations. This implies that social structure increasingly emerges out of interactions based on more individualistic behaviors, hence conditions that favor biological kin selection have become relatively less important. At every level among the chimpanzees, behavior is more individualistic and less constrained by environmental conditions (Boesch and Boesch-Achermann 2000). As a result, chimpanzee social structure has a form emergent from performance based on individualistic behavior, and the outcomes are less predictable in an *a priori* sense due to "the new fluid social environment" (Boesch and Boesch-Achermann 2000: 265).

Individualistic Behavior in the Chimpanzees

The individualistic behavior of chimpanzees (and other great apes) parallels that of humans: "The orangutan, gorilla and chimpanzee especially resemble man in this individualization of behaviour" (Yerkes 1927: 185). This theme has been repeated by more recent researchers, "Our chimpanzees show pronounced individuality" (Nissen 1956: 412).[2] In a similar vein, it has been noted that "the pattern of chimpanzee behavioural diversity seems to be more like that of humans than of other animals" (McGrew 2003b: 179). A recent review of primate behavior concludes, regarding the great apes, that "the most intriguing finding is the selection for high individuality ... apes are rather self-contained individuals" (Maryanski and Turner 1992: 30). By individualistic behavior we are concerned not with what might be called personality differences among individuals in a social group, but with differences in individual behavior affecting the outcomes of social interactions relevant to the structure, organization and coherence of a social group.

The social complexity of a group relates to the degree to which group coherency through time depends on each actor in a social context taking into account the actions and behaviors of other actors in the group (de Waal and Tyack 2003). The diversity of such actions and behaviors need not be dependent, though, on the group size per se. If complexity did relate directly to group size, then a flock of starlings with tens of thousands of birds would be an extremely complex group. However, a flock of birds, a school of fish or a herd of antelope has coherency without being socially complex because "simple rules of interaction among the individuals are sufficient to produce collective behavior" (Ballerini et al. 2008: 1232). For starling flocks, measurements show that each bird pays attention only to the six birds closest to it, then matches their velocities and is attracted towards or moves away from them to avoid collisions (Ballerini et al. 2008). From the perspective of the individual starling, the social world of the flock is simple: act according to the average velocity and location of the six birds nearest to you and ignore other birds and other behaviors. Consequently, we can consider the flock to be composed of behaviorally equivalent individuals—the birds nearest to a given bird may change, but the bird behaviors responsible for the patterned, collective behavior of a flock remain the same. Hence the effective group size for measuring the complexity of the flock is $n = 1$, corresponding to the average behavior to which an individual bird responds, and the relevant set of behaviors or attributes from which the collective, flock behavior emerges is velocity and distance. The flock, despite its seeming complexity of behavior as a collectivity, is a socially simple society.

Measure of Social Complexity

We can measure the complexity of a social group, to a first approximation, by the maximum size for a cohort of individuals in which each of the individuals in the cohort has a distinct set of behaviors that can lead to different outcomes, depending on which of the individuals in the cohort is involved in social interactions. Human societies, as will be discussed below, have accommodated the social complexity introduced by individuality of behaviors through cultural constructs such as kinship systems that transform interaction between individuals to one of acting in accordance with culturally constructed roles. When interacting through mutually understood roles, social complexity is determined by the number of distinct roles and not the number of actors.

Absent cultural systems for transforming interactions between individuals into interactions based on mutually understood roles, individuality in behavior implies that a predictable outcome for the social interaction by individual A with individual B depends on prior face-to-face interaction by A with B and cannot be predicted by A from experience gained through interaction with another individual C. Among the primates, the lemurs have simple social groups in this sense since, for example, the "ring-tailed lemurs express *only minor behavioral differences between individuals* ... [with] no extensive differences among individuals as would be seen among anthropoid primates" (Boyd 2000: 39, emphasis added), and "relationships between any two lemurs are remarkably black or white: either almost wholly affiliative or almost wholly antagonistic" (Jolly 1998: 5, with references to Kappeler 1993a, 1993b). Hence what a lemur learns or understands about social interaction with one or two lemurs can be generalized to interactions with other lemurs. At the other extreme, chimpanzees have complex social groups due to the individuality of chimpanzees that leads both to different behavior outcomes by individuals in social interactions, such as differences in the "style" of alpha males (Foster et al. 2008) and differences in coalition strategies (de Waal 1982; Nishida and Hosaka 1996). An increase in individuality, even absent any change in group size, will affect the social complexity of a group. Social complexity measured by the effect the degree of individuality of group members has on different outcomes for social interaction, including interaction by individuals with dyads composed of coalitions or alliances, increases non-linearly with the square of the number of individualistic group members (see Figure 4.3).

Group size per se is not, then, a sufficient measure of social complexity. Although social groups for the terrestrial, ring-tailed lemur (*Lemur catta*) are composed of around fifteen individuals (Jolly 1966), from a social complexity viewpoint the number of individualistically distinct group members is $n \approx 2$, hence we can characterize the ring-tailed lemur as having groups with a simple social structure. We find corroboration for this assessment of the simplicity of lemur social groups (in comparison to the anthropoid primates) in their mean neocortex ratio < 1.20 (Kudo and Dunbar 2001), a value significantly less than the neocortex ratio for OW monkeys (see Figures 2.8, 4.1 and 4.4),

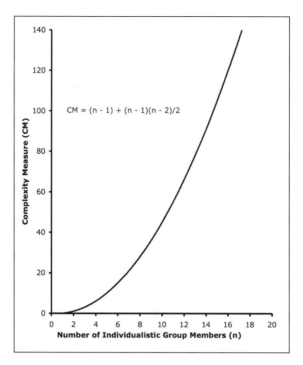

Figure 4.3 Graph of number of individualistic group members versus social complexity measured as the number of individualistic group members plus the numbers of dyads involving individualistic group members but excluding the reference individual (CM = (n-1) + (n-2)(n-1)/2).

despite the relatively small change in the size of social groups between the phylogenetic levels corresponding to the ring-tailed lemur and the arboreal OW monkeys.

Social Organization and Social Complexity

As we can also see from Figure 4.1, the slope for the increase in the neo-cortex ratio between the lemurs and the arboreal OW monkeys (dashed line) is much steeper than the slope for the change in the neocortex ratio going from the arboreal OW monkeys through the terrestrial OW monkeys and then to *Pan* (solid line). The difference in these two slopes relates to a major change in social organization when going from a group composed of similarly behaved group members to a troop structure composed of increasingly individualistic troop members subdivided into matrilineal groups and organized according to a linear dominance hier-archy (see Figure 4.4). The latter implies, as discussed above, that troop

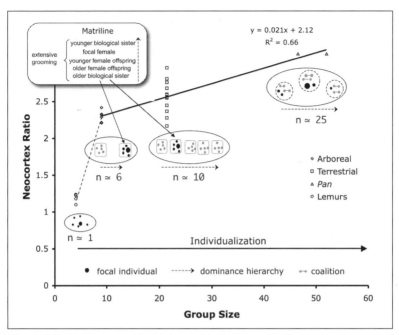

Figure 4.4 Neocortex ratios, schematic diagrams for the group, troop and community structures of the lemurs/prosimians, OW monkeys, and chimpanzees, respectively, and individualization of behavior dimension. The lemur group structure has social complexity for a focal individual with $n \simeq 1$ due to similarity of behaviors among group members. The organizational change between lemurs/prosimians and OW monkeys, along with an increase in individualization of behavior, is reflected in the increase in the neocortex ratio. The OW monkey troop structure is based on a linear hierarchy of matrilineal units structured as indicated for a matriline. A female focal individual has a social complexity of approximately $n \simeq 6 - 10$ based on the number of individualistic females in a matriline and the number of matrilineal units in the linear dominance hierarchy. The transition from OW monkeys to chimpanzees includes change from a matriline troop structure to a male fission-fusion community with short term social units. The male linear dominance hierarchy cross-cuts social units. Coalitions and alliances are central to the social environment. The social complexity for a focal individual is around $n \simeq 25$, the number of male dyads in a social unit of 6 males plus the number of males in the community. Absent small social units, the complexity for a focal individual would be approximately $n \simeq 125$, an order of magnitude more complex than the social complexity of OW monkeys. The social units reduce the complexity to around $n \simeq 25$, consistent with the change in the neocortex ratio.

size can increase with relatively little effect on the social complexity of the troop. This can be seen in the moderate slope for the increase in the neocortex ratio when going from the arboreal to the terrestrial OW

The Chimpanzee Middle Point

monkeys. The matrilineal structure makes possible both greater individuality and larger group size without the social domain becoming too complex. Selection for individuality, it should be noted, is antithetical to biological kin selection, as the latter introduces genetically based social behaviors that, by the nature of biological kin selection, will be distributed over the members of the cohort of genetically related individuals in a manner subject to biological kin selection. This will lead to introducing homogeneity of behaviors across biologically kin related individuals, whereas the introduction of individuality does the reverse.

The transition from the troop structure of the OW monkeys to the community structure of chimpanzees continues the moderate trend in the relationship between the neocortex ratio and social complexity (solid line in Figure 4.4). We do not find a strong upward trend for the neocortex ratio in comparison to group size as might be expected given that a community form of organization without social subunits and with individualistic members would create a highly complex social environment (see Figure 4.4). Instead, the fission-fusion social structure bounds the potential complexity by focusing social interaction among males within temporary, small social units. The social units differ from the matrilineal units of the OW monkeys by being neither internally stable nor structured externally through a linear dominance hierarchy. The organization of the social units appears, instead, to be an adaptation to the increased individuality of chimpanzees. As individuality increases, the size of a stable social unit decreases and social behaviors are less focused on biological kin, so conditions favoring the introduction of social behaviors that promote social coherency through biological kin selection are reduced. In place of biological kin-based social behaviors, the coherency of social units depends increasingly on extensive, face-to-face interaction, but the efficacy of so doing, for a fixed size of social units, decreases rapidly with increased individuality. As a consequence, social cohesion is maintained primarily by reducing the size of social units.

Selection for substantial individuality in the great apes appears to have driven their forms of social organization to limiting cases—to solitary behavior in *Pongo*, to single male harems in *Gorilla*, and to small, unstable social units in *Pan*. The social units in *Pan* communities have a size no longer bounded externally by factors such as the density and distribution of food resources and internally by internal organization leading to a stable dominance hierarchy as happens with the OW monkeys. Rather, the social units are formed according to the internal dynamics of interaction among individualistic chimpanzees. This leads to a social system based on "a network of positions of influence" (de Waal 1998: 207). This horizontally structured network cannot be subsumed within a vertical dominance hierarchy among dominant males (de Waal 1998).

From Chimpanzees to Humans

The social organization of chimpanzees appears to be at the limit for what is possible with biological processes such as biological kin selection as a way to introduce social behaviors that provide social coherency in response to the socially destabilizing effects of increased individuality. Factors such as selection for social intelligence make possible, on the one hand, enhanced ability to engage in social responses that can integrate the wider range of behaviors introduced by increased individuality. On the other hand, the same selection has the potential of increasing individuality. With increased individuality, there should be increased selection for general, rather than specific, behaviors such as the capacity for social learning rather than specific social behaviors. With the OW monkeys, the specific behavior of repeated interventions by the biological mother in agonistic encounters of her daughter with other females seems to be the catalyst for the dominance rank of her daughter, but establishing dominance relations among chimpanzee males does not depend on any specific behavior. Rather, dominance relations arise from a wide range of behaviors varying from sheer physical strength to strategic grooming to formation of support coalitions. With their increased individuality, chimpanzees appear to engage in a social world that is inherently less constrained in an ostensive sense and thus more subject to the emergence of social structure through performative social interaction than is the case with the OW monkeys.

Increased Individuality and Social Complexity

The trend from the OW monkeys to the chimpanzees is, therefore, one of increased individuality coupled with a shift from ostensive to performative based social organization (left side of Figure 4.5). Only the increased individuality carries forward to hunter-gatherer societies. The trend towards performative forms of social organization does not. As discussed in chapter 3, hunter-gatherer societies are constructed around culturally framed patterns of behavior, especially behavior based on cultural kinship, and are thereby more ostensive in the form of their social organization but for different reasons than is the case for the OW monkeys.

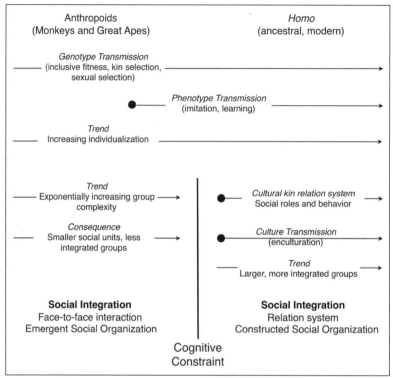

Figure 4.5 Cognitive and behavioral constraint between anthropoids and *Homo*. Constraint circumvented by shift from learned to conceptual social relations, with the latter leading to cultural kinship systems. Reproductive and imitation transmission comprise dual inheritance. Enculturation transmission is the process by which cultural systems are transmitted across generations. Solid circle indicates the start of a new organizational dimension.

The increased individuality of the chimpanzees both acts against social coherence and places greater demand for social learning through face-to-face interaction among the members of a social unit. The lack of stability of male social units implies that males are repeatedly faced with a changing social environment, one in which they are constructing, as noted by de Waal, "a network of positions of influence." This has the effect that selection for increased capacity for social learning leads to chimpanzee communities becoming more diverse. Examples of communities exhibiting instances of the rudiments of culture, defined as the social transmission of behavior traits, become increasingly prevalent. Differences among chimpanzee communities in behaviors socially transmitted through imitation include dozens of examples ranging from tool usage to grooming to courtship behaviors (Whiten et al. 1999: Table 1), with each of these behaviors distributed in a community-specific manner. These examples attest to the individuality of chimpanzee behavior and to the degree to which chimpanzees are capable of novel behaviors generated and transmitted directly at the phenotypic level rather than genetically at the genotypic level (see phenotype transmission, Figure 4.5). This also underscores the increased importance of individual learning and imitation of the behaviors exhibited by other community members, and it expands the breadth of the behavioral repertoire of the chimpanzees in a community.

Behavior Traditions Versus Rule-Based Behavior

The socially transmitted behaviors make it evident that we share with the chimpanzees not only a substantial part of our genome but also the capacity for generating socially transmitted behavior traditions. The behavior traditions are not, however, the equivalent of human cultural systems, and the presence of behavior traditions is not the marker of what is meant by culturally constituted societies. As Lévi-Strauss commented, "Wherever there are rules we know for certain that the cultural stage has been reached" (Lévi-Strauss 1969[1949]: 8). With this remark, Lévi-Strauss identified that it is a system of conceptual relations held in common— expressed as rules—and giving structure to the way individuals are conceptually linked to each other in a mutually understood manner, thereby forming a "complex whole," that constitutes what we mean by culture. Cultural systems are not reducible to an ensemble of behaviors or relations distinguished simply because they are learned through experience and transmitted in a social context. It is not the mode of transmission and its consequences for the frequency distribution of traits that is critical. Rather, what constitutes culture systems is systematically related ensembles of mutually understood concepts that provide a framework within which behavior takes place.

Individual Concepts Versus Conceptual Systems

The kinship terminology systems we discussed in chapter 3 for hunter-gatherer societies make this distinction between a collection of individual concepts and a system of concepts evident. At one level, a kinship terminology can be considered to be a list of words that label categories of kin produced through sexual reproduction and identified as kin through genealogical tracing and/or marriage ties. That the kin terms are transmitted in a social context as part of a child becoming enculturated is evident. However, what a child learns through enculturation is not simply a list of words that we refer to as kin terms. If the traits being transmitted from one individual to another were just individual words with meanings that make them kin terms derived from the categories they label, then it should be evident to us when a word is a kin term or not. Yet by what criteria do we know when a word is a kin term and therefore that the referent of that word is a kinsman? Does the use of *father* as in Father Junipero Serra, the founder of the California missions, make him a kinsman? Obviously not, but on what basis? What about the word "relative"? Is that a kin term collectively referring to all of one's kin? Or consider the use of the terms "uncle" or "aunt" by children for close friends of their parents in American society. Does this mean the close friends of one's parents are also kin? What about the variants in reproduction made possible through medical advances? If a woman has another woman's egg fertilized through artificial insemination and implanted in her uterus and then gives birth to a child, who is the mother and who is the father of that child? Answers to questions like these are not derived by reference to the definition of kin terms considered in isolation, but through the manner in which there is an ensemble of conceptual relations, expressed through our kin terms, that is central to our assessment of whether we are using a word in a kin term sense or not.

As discussed in chapter 3, a kinship terminology goes beyond simply being a list of words relating to kin. What makes it more than a list of words is the fact that a kinship terminology is a system of concepts and is not constructed in an ad hoc manner in response to external exigencies (that is, it is not the institutionalization of already existing patterns of behavior), but in a way that makes the kinship terminology as a whole into a logically consistent, coherent system of concepts. Logical consistency makes it possible for the terminology to be shared conceptually, not in the sense of individual bits of knowledge transmitted and remembered each generation, but in the sense that the logic of the system of kin term concepts ensures that all societal members understand the system of concepts in essentially the same way. The terminology can be viewed as a system of symbols—the symbols being the kin terms—that stand for the concepts and ideas we have about kinship relations, and the logic of the terminology enables us to do computations with kinship relations. In this sense a kinship terminology system is analogous to our natural number system. Our natural number systems consists of a system of symbols, 1, 2, 3, ..., that represent our concepts about counting quantities and conceptually what it means to additively combine counting quantities. These counting concepts have a logic to them that we express through our system of arithmetic based on doing computations with number symbols. The system of symbols ensures that every competent user understands that $1 + 1 = 2$, $1 + 2 = 3$, $2 + 2 = 4$, and so on. When we use the natural numbers and do symbolic computations with them, we "know" that the computations made by one competent person will be the same as the computations made by another competent person using the same number symbols. When errors occur, such as computing $2 + 2 = 5$ erroneously, the error can, in principle, be discovered by any competent user and corrected, with everyone agreeing on the correction.

A kinship terminology plays an analogous role in kinship systems. The users of the American kinship terminology "know" that if someone is your father, then his son is your brother and his sister is your aunt and when you refer to someone as uncle in a kinship sense and that person refers to someone else as daughter, then you refer to the latter person as cousin. It follows (to answer our previous questions about whether words are being used to express kinship concepts) that Father Junipero Serra is not "my father" in this sense (though he might be "my father" in the context of Catholic idea systems about priests), and the child of a male friend of my parent is not my cousin and so my parent's friend is "my uncle" in only a metaphorical sense.

Two Hypotheses for the Transition to Hunter-Gatherer Forms of Social Organization

The transition from a chimpanzee-like ancestor to our hominin ancestors and their evolutionary development into *Homo sapiens* is not simply the extension of trends we find in the social systems of the non-human primates. That trend implies we should find the organization of human societies emerging from extensive, face-to-face interactions, much like the form of social organization in social units in chimpanzee societies arises from extensive interactions among males. The problems that occur with assuming a direct connection between a chimpanzee-like social system and the kind of societal organization we find in hunter-gatherer groups have been identified by Bernard Chapais (2008: 172). Instead of a direct connection, he suggests an indirect evolutionary pathway, He observes that the chimpanzee form of social organization lacks critical elements that are part of hunter-gatherer societies. Two critical differences relate to (1) tracing cultural kinship relations through both the mother and father positions that are part of family relations and (2) including individuals outside of one's residence group as part of one's cultural kin.

(1) Transition Through Mate-Guarding and Pair-Bonding

Chapais argues that both differences between the chimpanzee form of social organization and hunter-gatherer societies could arise from a social system based on pair-bonding, hence the critical transformation in his scenario is the transition from a multi-male, multi-female, chimpanzee-like social system to one that is based on pair-bonded males and females. He posits a two-step transition for this evolutionary pathway. First, there would be a transition from the multi-male, multi-female kind of organization found in the chimpanzees to a harem-like system in which a male controls sexual access to a cohort of females, much like the pattern found in modern-day gorillas or the hamadryas baboons. Second, there would be a transition from the harem-like system to a pair-bonded system, much like the pattern found in modern day gibbons and siamangs, though with a social unit based on a number of pair-bonded males and females rather than the gibbon and siamang pattern of a single mating pair and their offspring

As to how the transition would take place, he comments: "The exact circumstances that favor the evolution of multi-harem structures in extant nonhuman primates, and by extension in extinct hominids, remain to be further investigated by behavioral ecologists" (2008: 174) and attributes the transition to "ecologically constrained mate-guarding strategies" (175). For example, Chapais asserts that mate-guarding strategies would not be possible when females forage in large groups but might occur when they forage in small groups, and this difference in foraging pattern would depend on ecological conditions. What the ecological conditions would be and whether they characterized the habitats of the hominin ancestors is left open.[3] The second transition from a harem-like form of social organization to pair bonding would occur, he suggests, if previous variability in male-fighting ability as part of male-male competition were somehow equalized, thus making it impossible for one male to successfully guard a cohort of females against other males. He proposes that this equalization could have taken place through objects used as weapons: "Any tool, whether it was made of wood, bone, or stone ... could be used as a weapon ... especially [with] one that could be thrown some distance, any individual, even a physically weaker one, was in a position to seriously hurt stronger individuals. In such a context it should have become extremely costly for a male to monopolize several females" (177). For Chapais, then, the origin of pair-bonding—what he initially refers to as "stable breeding bonds" (25)—is a one-sided relationship in which a male actively controls sexual access to a female.

In addition to these two steps, Chapais has an implicit third step: transformation of asymmetrical pair-bonding from male mate-guarding to symmetrical pair-bonding predicated upon emotional ties between male and female. Chapais argues that human pair-bonding is biologically grounded and so there are "various physiological, neuro-biological, and emotional processes involved in the formation and maintenance of human pair-bonds" (162). However, a mate-guarding strategy does not, by itself, account for the emotional attachment involved in human pair-bonding, hence the implicit third step in his scenario. Further, the mate-guarding strategy does not appear to be necessary as a precursor for the development of pair-bonding through emotional attachment. We will return to this point below.

With stable breeding bonds and social units composed of several pairs of bonded adult males and females, offspring can differentiate their behavior when interacting with a male bonded to their biological mother from other males (and vice-versa). This, in combination with behavior already differentiated when interacting with their biological mother in comparison to other females, provides a behavior pattern that is the precursor, in his argument, to bilateral kinship. Coupled with the chimpanzee male philopatric pattern, males would have a continuing relationship with their biological sisters and daughters when females migrate to different communities as they become sexually mature. The continuing relationship, he asserts, would provide the foundation for non-aggressive interaction between communities, a key aspect of the change from non-human primate to human forms of social organization. Chapais also suggests that another consequence would be for in-laws (affinal relations) to become part of the social sphere of individuals, and through this argument he connects his argument with that of Lévi-Strauss and others regarding the role of exogamous marriages in establishing social ties between different groups in human societies.

Chapais's argument is intriguing as it brings together a number of seemingly disconnected pieces. Nonetheless, it makes two crucial assumptions that are problematic. One has to do with equating mate-guarding with the emotional ties of pair-bonding and assuming that mate-guarding/pair-bonding behavior is essentially the kind of behavior associated with marriage. The other has to do with the related assumption that cultural phenomena are essentially the institutionalization of already existing behavior patterns.

By equating a mate-guarding strategy with pair-bonding based on emotional attachment, Chapais has unnecessarily conflated two different relations that have different foundations and consequences. The problem with the conflation is especially evident in human societies. Human pair-bonding through emotional attachment centers on a dyadic relationship in which "a man or woman can retain sexual access to a desired partner only to the degree that he or she is able to remain more desirable than others who may seek access to that partner" (Bell 1997: 238), whereas mate-guarding is more akin to marriage in that for a group such as the Nuer in Africa (but not universally) "only a husband has an institutionalized, socially supported *right* to control her sexuality" (1997: 238, emphasis in the original). For a male to have, as Bell puts it, "an institutionalized, socially supported right," there must be other persons who confer those rights and provide the social support. Altogether, this makes relations derived through marriage of a fundamentally different sort than those derived through pair-bonding.

Further, although the quote regarding the Nuer can be read as suggesting similarity between mate-guarding and marriage, the similarity is superficial since marriage does not derive, as does mate-guarding, from the attributes of the individuals involved. Rather, marriage is a cultural institution that assigns the individuals concerned to "a particular placement in social space" that determines "the rights to various resources that apply to that position and the responsibilities that must be fulfilled in order to validate one's continued placement [in that social space] ... [and] the marital tie within the space of social relations is defined by a particular configuration of rights" (Bell 1997: 241). In brief, marriage does not exist as an institution outside of a social context and can neither be reduced to a mate-guarding strategy nor to pair-bonding, though it may (but need not) incorporate aspects of each of these.

Pair-bonding also contrasts with marriage by being activated through the male and female in question, and we can infer the degree of pair-bonding by the extent to which mating behavior is directed primarily towards one individual and over what time frame. Marriage, in contrast, has to do with specifying the interests of societal members with regard to the conditions under which sexual behavior may take place and the criteria by which a woman's offspring are accorded full status as members of a society. Whereas mating is the means by which offspring are produced, marriage is the means by which offspring are given full status as societal members and can, under specified conditions, incorporate mating with individuals other than the two individuals who are the subject of a marriage event. It can also involve two individuals of the same sex, and so no mating will occur between them. Mating rights, lack thereof, or the conditions under which they may be invoked are expressed through marriage, but marriage is not simply mate-guarding or pair-bonding writ large.

In the argument to be presented in the next chapter, the evolution of emotional attachment in pair-bonding will be seen to occur as part of male provisioning of a female and her increasing control of his sexual access to her. The change in behaviors, it will be argued, came about as meat consumption became an increasingly important part of the hominin diet and, simultaneously, females had less access to meat resources due to decreased mobility induced by offspring with increasingly reduced motor and neurological development at time of birth. In this scenario, pair-bonding still plays the role envisaged by Chapais with regard to identification of the biological father, but pair-bonding does not become the prime mover for the transition from primate to human forms of social organization. Instead, the transition arises through a fundamental change in how relations among individuals in a social unit are established, which Chapais does not discuss.

The second, critical assumption in the scenario presented by Chapais posits a biological underpinning to the behaviors found in hunter-gatherer societies, with culture largely playing an ancillary role. In Chapais's argument, pair-bonding provides the context in which biological kin selection can act on biological kin related to an individual through biological mother and/or biological father. However, if the behaviors in a hunter-gatherer society can be achieved by invoking nothing more than the kinds of arguments used to account for patterns of social organization among the non-human primate species, there is no reason why a cultural kinship system, especially in the sense of a complex whole as discussed in chapter 3, should have arisen as the basis for social organization. Instead, what we are calling cultural kinship would mainly serve to institutionalize already existing behaviors, as Chapais argues is the case for unilineal descent systems: "This type of organization thrived in various primitive forms *before it was institutionalized*" (2008: 275, emphasis added).

Under his scenario, the formation of a cultural kinship system becomes the analogue of the Baldwin effect for biological systems.

Towards the end of the nineteenth century, James Baldwin argued that behaviors learned or otherwise introduced into the behavioral repertoire of organisms at a phenotypic level would, if the same pattern continued over sufficiently long periods of time, be co-opted, as it were, at a genetic level through random mutations eventually leading to genetic variants that provide a biological basis for the behaviors in question at the phenotypic level. Baldwin argued that it would be less costly and less prone to error for behaviors otherwise individually learned to become transmitted genetically and as part of the phenotype through biological development rather than through individual learning.

Chapais makes a similar argument except that culture and culture transmission have replaced genes and genetic transmission. Chapais follows in the tradition of those who argue that behaviors initially learned individually are eventually co-opted, or institutionalized, through culture, and that subsequently the behaviors are transmitted through cultural transmission rather than individual learning. His argument has the following difficulty. Culture can be no more organized or constitute a more logically structured system than the behaviors in question. However, cultural systems such as kinship terminologies are logically structured, and differences in the logical structure of kinship terminologies discussed in chapter 3 do not correspond to structurally similar differences in behaviors. As already discussed, the kinship systems of hunter-gatherer societies have to do with constructed, ostensively formed systems of cultural kin, not derived, performative systems of biological kin behavior. Thus the assumption that cultural phenomena are essentially the institutionalization of already existing patterns of behavior is contradicted by the demonstrated properties of those cultural phenomena.

(2) Transition Through Loss of Dominance Hierarchies and Normative Behavior

An alternative pathway to human social systems exemplified by hunter-gatherer societies has been developed recently by Benoît Dubreuil (2010). Dubreuil's pathway also involves two steps: first, the loss of dominance hierarchies as the basis of social structure and second, neurological changes through encephalization that enabled norm-based behavior and social cooperation in areas of mating and reproduction. The loss of a dominance hierarchy is said to be triggered by the development of cooperative feeding by our ancestral hominin species, *Homo erectus*, and so dates back to around 1.5 million years BP. This trend, he argues, was enhanced through the introduction of cooperative breeding in which a biological mother received parenting assistance from other group members. Cooperative breeding then purportedly arose along with more pronounced encephalization in our ancestral lineage during the Middle Pleistocene, around 500,000 years BP, and is associated with our ancestral forms classified as *Homo heidelbergensis*. Dubreuil rejects Chapais's second step of a shift to pair-bonding through weapons making it possible to kill competitor males at little cost as it would require, he argues, just the right balance of lethality for pair-bonding to be a stable equilibrium arising from the development of weapons. If a male could easily kill any competitor, he argues, then it would be in his interest to kill all competitors, and the only stable solution would be a single male mating with all females; that is, there would be a shift to a large-scale, *harem* system without male competitors. If the cost was too high, the shift would only be to a reduced *harem* system, not to monogamous pair-bonding.

For the first step, Dubreuil draws upon the argument made by Christopher Boehm (1999) regarding the egalitarian character of hunter-gather and nomadic tribal societies discussed in chapter 3. According to Boehm, the egalitarian character derives from an egalitarian ethos that reverses a natural tendency to dominate other individuals. Boehm centers his argument on ethnographic evidence showing that when an individual in a hunter-gatherer society attempts to parlay his particular skills, such as being an excellent hunter, into a dominance relation vis-a-vis other males, he is countered with active resistance on their part. Their resistance can vary from ridicule or gossip, to social ostracization, and to killing in extreme cases. Other males are acting, he argues, according to an ethos of equality among males that essentially denies to any individual the right to dominate others or to expect subservience from them. Another way to phrase it would be to say that dominance can arise when a male recognizes himself as an individual who can express his individualistic behavior in the manner he sees fit and acts according to his capabilities. However, a stable dominance hierarchy, by its very nature, is based on the willingness or predisposition of individuals to act, instead, according to their position in the dominance hierarchy.

For the OW monkeys, the assertion that the behaviors leading to a stable dominance hierarchy arose through biological kin selection implies there is selection for non-individualistic behavior; that is, there has been selection for biologically based behaviors that are enacted in accordance with the dominance hierarchy. As we have seen, forming a stable dominance hierarchy, along with behaviors predisposed towards conforming to the hierarchy, reduces the social complexity of a group over what would happen if all members of the group acted independently. In the reverse direction, the stability of a dominance hierarchy is challenged if group members become more individualistic in their behaviors, and with increased individualistic behaviors the dominance hierarchy must increasingly be maintained by force or other means whereby one individual can maintain control or power over other individuals. Under these conditions the dominance hierarchy may lose its previous stability once there is the ability to form coalitions, since a more powerful, dominant individual can be overcome by coalitions of less powerful individuals. We see this change to less stable dominance hierarchies arising from coalition formation by males among chimpanzees.

That dominance hierarchies of the kind found in non-human primate societies disappeared during the evolution to modern *Homo sapiens* is evident from extensive data on modern hunter-gatherer groups (Boehm 1999). Less evident is whether the egalitarian nature of hunter-gatherer societies can be reduced to the presence of a culturally constructed "egalitarian ethos" (Boehm 1999: 66), even if coupled with "cognitive mechanisms related to social norm and sanction" (Dubreuil 2010: 50). The importance of social norms for implementing a purported egalitarian ethos relates to the fact that humans, but not non-human primates, appear to have a psychological and neurological basis for both norm following and sanctioning violators of norms. Humans experience negative emotions such as fear, shame, guilt and disgust when violating a social norm and emotions such as anger and indignation when seeing the violation of a social norm. These emotions motivate both internal constraint and sanctioning behavior (Dubreuil 2010, and references therein). The egalitarian ethos, then, is said to express the culturally negative interpretation of behaviors used by an individual to establish a superordinate relation concerning another group member through the idiom of normative behavior. In this framework, the behaviors initiated in response to the emotions associated with instances of norm violation are seen as leading to the implementation of that ethos. Left unanswered, though, is the source for, and content of, the social norms.

The norms we see in hunter-gatherer societies are not specific maxims for behavior—such as "cooperate with others" or "share an animal killed through hunting"—but are part of cultural constructions specifying those with whom cooperation is expected or those who are being allocated the right to make claims on a resource. Cooperation is typically expected among close cultural kin as indicated by Fortes's kinship Axiom of Amity. Similarly, sharing of meat is based on culturally constructed collective ownership of a killed animal and is implemented by cultural rules according to which collective ownership becomes transformed into individual ownership, not norms about specific behaviors. The norms related to the hypothesized egalitarian ethos derive from these shared cultural systems and are thus part of a culturally constituted universe within which behavior by group members takes place. Trying to make sense of human behavior without reference to this broader, culturally constituted universe is a chimera. It is as if we tried to explicate the motion of the celestial objects by measurements made on their time-based location but without taking into account the gravitational force that is part of the space-time universe within which those objects are embedded. The equivalent of the gravitational process is culturally based information processing.

Distinction Between Biological and Cultural Bases for Information Processing

From an information processing perspective, it appears that our brain incorporates both a cognitive processing system derived through evolution as part of our biological heritage and a cultural processing system that came into play during the evolutionary development of modern *Homo sapiens* (Read 2010a). Both of these systems operate on information from external events that is then internalized through our sensory system. Each provides mental representations of external events according to the properties of the two information processing systems. Our biologically and culturally based mental representations, in conjunction with internalized goals and strategies, lead to possible behaviors through either our biological based cognitive processing system or our culturally based processing system. When these possible behaviors are not the same, a decision process adjudicates between them, taking into account the context and conditions in which we are currently engaged. The result is the specific behavior that we engage in.

A simple example of this process occurs when we walk across a flat grassy area bounded by sidewalks laid out in a rectangular fashion. The cultural interpretation of the sidewalks is that they define the walkway that should be followed by a pedestrian when walking across that area to go from point A to point B. The cognitive interpretation based on our biological heritage of following paths that minimize time and energy leads to a straight-line path from A to B, independent of the sidewalks. As we approach point A, we typically make a non-conscious decision to either follow the sidewalk from A to B or to walk across the grass. Some individuals follow the sidewalks and others the straight path across the grass. The same individual on one occasion may follow the sidewalk and on another occasion the straight path. This variation both between and within individuals indicates that there is no single best choice between a path that follows the sidewalk and one that goes across the grass. Instead, two possible behaviors are available to an individual, one determined through biological processing that leads to a least effort path and the other through cultural processing that leads to a path that is culturally marked by the sidewalks and is based on our shared understanding of the meaning of the sidewalks for pedestrians. The decision as to whether he uses one behavior or the other depends on parameter values from the context in which the pedestrian is acting. On one day the pedestrian may be guided by what others are doing, and if they are walking on the sidewalk, do the same. On another day, the same person may be late for an appointment and, eager to minimize time, walk across the grass.

We can distinguish between a behavior derived through biological processing versus one derived through cultural processing according to whether the behavior can be accounted for by reference just to traits and parameter values for the individual in question or whether it also requires reference to a larger system that involves other individuals. We can account for an individual following the least time and energy path by assuming the individual has, as a genetic trait, a cognitive and visual processing system that constantly orients the direction of moving towards the location of the goal. The frequency of this trait across other individuals, though relevant for assessing possible evolutionary change in a population, is not a determinant of the behavior of the individual in question.[4] Either the individual has the genetic trait or does not. In contrast, when we see the same individual following the sidewalk, we cannot account for the behavior just by referring to a normative cultural trait such as "follow the culturally specified walkway" since we do not know what constitutes the "culturally specified walkway" without knowledge about the cultural system of walkways for which the sidewalk is an instance. When the pedestrian is observed following the sidewalk for cultural reasons, the behavior makes sense to those who share with the pedestrian the cultural meaning of sidewalks and grassy areas. However, absent that shared cultural meaning, the observer, say a person transposed from a traditional hunter-gatherer society, would likely impose her or his cultural framework for interpreting the meaning of grassy areas with sidewalks, or may not even recognize that there is a meaning associated with those features.

Shared Cultural Meaning

Shared cultural meaning in this sense means more than just the statistical frequency distribution of a particular meaning across a population of individuals; it is the meaning held in common and derived from the cultural idea system transmitted through enculturation. It is the cultural idea system that enables an individual, when meeting with a novel situation, to understand what is a walkway. Suppose we arrive at a new area with plants other than grass and with sidewalks that are curved, hence not simply a scaled up or scaled down replica of the grassy area with sidewalks laid out in a rectangular manner. We still make the interpretation that the sidewalks are the culturally appropriate walkway. We do this by our understanding, through enculturation, that flat, level bands of concrete with a surface at about the same height as the ground and with a width that would accommodate one to several persons are designed to be functional as a walkway, and that this functionality does not depend on whether the concrete band is straight or curved. Further, we understand through enculturation that the non-sidewalk areas are not walkways and are usually designed to have a functionality other than being a walkway, such as aesthetic appearance, hence they may have a variety of plants and not just grass. It is in this sense that understanding the behavior of the person who uses the sidewalk as a path requires reference not just to a phenotypic trait the person may have obtained through, say, an imitation process, but to the cultural idea system in which that trait is an instance and how the cultural idea system relates to the functionality of sidewalks expressed through the design of the areas consisting of both plants and sidewalks.

Involved here is the distinction between shared meanings due, for example, to an imitation process for direct, phenotypic transfer of a behavior from one individual to another and the indirect transfer of a behavior through transmission of cultural idea systems through enculturation. The distinction is analogous, at the genetic level, to the difference between direct transmittal of a gene from one organism to another (a form of gene transfer that occurs among bacteria) versus the transmittal of a genetic system through reproduction. The latter entails the former, but the biological consequences of transmission through reproduction are not reducible to simply the sum of a series of individual gene transfers. Minimally, transmittal of the genetic system also relates to the structural organization of genes and other aspects of the genome and how they affect the phenotypic development of an organism. The same distinction applies to cultural traits that are part of cultural idea systems.

Cultural Transmission Through Enculturation

The distinction implies, then, that shared meaning for cultural traits has to do, more generally, with the transmission of cultural idea systems through enculturation (see middle right, Figure 4.5) and not just the frequency distribution of a cultural trait arising from phenotype-to-phenotype horizontal trait transmission through a cognitive imitation process (see upper left, figure 4.5). The latter refers to the frequency distribution over the population within which trait transmission takes place, and the former has scope determined by the boundary for individuals enculturated within the same cultural idea system. This difference can be illustrated with the hand-clasping used by the Mahale chimpanzees while grooming (see Figure 1.1). Though we do not know how an individual chimpanzee perceives the act of hand-clasping, evidently there is no equivalent to a cultural idea system transmitted through enculturation that underlines the behavior of hand-clasping and grooming. Consequently, whatever might be the meaning of the behavior (if any) to an individual chimpanzee and regardless of the extent to which individual chimpanzees may happen to perceive the behavior in the same way, there is no shared meaning in this behavior derived from transmission of a cultural idea system through enculturation. At most there is shared behavior arising, it appears, from a cognitive processing system that incorporates the imitation of behaviors as its output (lower part of Figure 4.6). Hand-clasping and other examples of shared behaviors by chimpanzees lack connection with the shared meanings transmitted through cultural idea systems that we find in hunter-gatherer and all other human societies. Consequently, the cultural processing box would have to be dropped in order for Figure 4.6 to model chimpanzee behavior. Rather than there being a continuous trend connecting the social structure and organization of common ancestry of the OW monkeys with the great apes, and then their social structure and organization with those of modern *Homo sapiens*, we have a hiatus as shown in Figure 4.5 and the eventual addition of information processing through a cultural component along with the biological and cognitive information processing as shown in Figure 4.6.

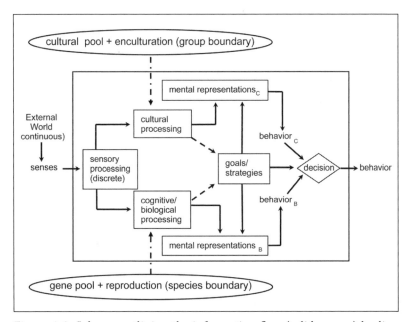

Figure 4.6 Schema outlining the information flow (solid arrows) leading from external stimulus via either cognitive/biological processing or cultural processing to behavior. Sensory processing transforms continuous stimuli into discrete segments, and these are the inputs for both our cognitive/biological processing systems derived from our biological heritage and our cultural processing systems that developed during the evolution of modern *Homo sapiens*. Each provides a series of mental representations that, in conjunction with internal goals and strategies, give rise to both a behavior$_C$ in accordance with the cultural interpretation and meaning of an external event and to a behavior$_B$ in accordance with a cognitive/biological interpretation and meaning of an external event. Adjutication between the two possible behaviors takes place through a decision making process that weighs whether the context and circumstance warrant a culturally or a cognitive/biologically directed behavior. The properties of the cognitive/biological processing systems are genetically determined, with a species boundary based on viable sexual reproduction. The cultural processing systems have a group boundary based on those enculturated in the same cultural systems. The dashed arrows indicate that goals and strategies are updated both by inputs from cognitive/biological processing and from cultural processing of inputs from external events. *Modified from Read 2010a: Figure 2.*

The hiatus is the consequence of a biological constraint arising from individualistic behavior replacing social behaviors introduced through biological kin selection. The biological constraint was only overcome by a new mode of social organization based on constructed kinship relations among group members rather than face-to-face interaction. The transition has been from social systems based on biological kinship integrated with face-to-face interaction to social systems based on cultural kinship in which social behavior is expressed through social roles associated with cultural kin relations. Americans, for example, not only have kin relations such as father, mother, brother or sister, they also know, or think they know (though neither necessarily with complete agreement nor with the same level of competence) how to be fathers, mothers, brothers or sisters or what behavior is expected by others when taking on one of these roles. The key piece for this transition lies in the evolution of cultural kin systems predicated on a system of cultural kin relations along with transmission of the conceptual system of cultural kin relations through enculturation. I will address the development of this new mode of social organization in the next chapter by modeling the evolution of cultural kinship systems that are central to the social organization of hunter-gatherer societies discussed in chapter 3.

Summary

Chimpanzee social organization appears to have reached a limit for the incorporation of individualistic behavior incorporated into biologically based social systems founded on face-to-face interaction. Along with individualistic behavior have come socially transmitted behaviors and community-specific patterns of behavior. While individualistic behaviors imply that chimpanzees have evolved some of the same characteristics of our behaviors, chimpanzee communities have not made the shift to forms of organization based on "culture as a complex whole." It is what is transmitted through social learning, not the mode of transmission per se, that is critical for comparing chimpanzee to human systems of social structure and organization. The difference is highlighted through the way a kinship terminology is a system of concepts and is not just a list of kin terms. A kinship terminology is shared conceptually because of its logical structure that ensures coordination in what is transmitted from one individual to another through enculturation. The terminology is a system of symbols through which kinship relations may be computed. It thereby defines who are one's kin and how kin relations among individuals may be computed. In brief, the terminology provides the conceptual basis for the structure and organization of the domain of culturally defined kin. Cultural idea systems, like the kinship terminology system, are information processing systems but differ from the cognitive systems that are part of the biological heritage of our species. We need to distinguish between behaviors made according to biological/cognitive information processing and cultural information processing. Shared cultural meaning is the outcome of cultural information processing among those individuals enculturated with the same cultural idea systems. A cultural idea system does not operate in isolation but as part of a system of interacting individuals with common enculturation. This implies that we need to distinguish between sharing due to transmittal of a cultural idea system through enculturation and sharing that just represents the statistical frequency of a behavior transmitted through social learning. The former is central to our understanding of the nature of human societies.

Notes

1 The group size is less than the 147.8 persons computed by Dunbar (1993) due to using a more detailed comparison of the neocortex ratio to group size in the non-human primates as shown in Figure 4.1.

2 Nissen goes on to comment, "It seems a safe assumption that both factors [genetics and experience] are effective in producing the individuality of chimpanzee behavior" (1956: 412). His assumption has been corroborated recently: "Genetic differences cannot be excluded as playing a major role in structuring patterns of behavioural variation among chimpanzee groups" (Langergraber et al. 2011: 415, but see also Lycett, Collard, and McGrew 2010). This does not negate the individuality of chimpanzee behavior, but simply underscores the way in which what we refer to as individual behavior is a complex mixture of both *sui generis* and learned or imitated behaviors, as well as behaviors with a genetic foundation (Read 2010a).

3 Foraging in small groups does not appear to be a sufficient condition for establishing a *harem*-like form of social organization since chimpanzee females forage in small groups but males (Pt or Pp) do not employ mate-guarding strategies. Whether early hominins foraged in small or large groups is unknown, and so we have no data on whether a *harem*-like form of social organization ever characterized, or was even feasible for, the ancestral species of *Homo sapiens*.

4 With herd or crowd behavior, the overall movement takes into account the behavior of other individuals. Locally a similar process operates, but with a time-dependent goal determined by a few neighbors as discussed in this chapter. At each moment, the behavior can be accounted for by the current goal without taking into account the behavior of other individuals.

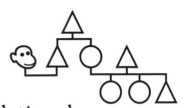

Transition to Relational
Systems of Social Organization

The evolutionary trajectory leading from the last common ancestor with the non-human primates to modern *Homo sapiens* diverged from an ancestral pattern of social organization based on social learning developed through extensive face-to-face interaction. This divergence, it will be now be argued, arose from a shift from experiential to relation-based social behaviors. Relation-based behaviors depend on categorization of dyads according to a relation recognized between the individuals making up each dyad (such as the category of biological mother-offspring dyads recognized by the macaques [Dasser 1988a, b] to be discussed below) and not on the traits of individuals per se (such as a category of aggressive males). Unlike trait based categorizations, relations between individuals make possible the formation of new relation categories through the "product" of relations; that is, categorizations constructed from computing the relation of a relation. This made possible social systems freed from emerging primarily out of patterns of social interaction learned through experience.

Whereas face-to-face interaction is a central component of ongoing and stable social behavior in non-human primates, ethnographic field-work shows, as discussed in chapter 3, that small-scale hunter-gatherer societies are structured around relations among individuals determined through a culturally constructed system of cultural kin relations that does not depend on prior face-to-face interactions. Instead, cultural kin relations both define and structure the domain of cultural kin for the members of a society and are linguistically expressed and communicated through the kin terms making up a kinship terminology. The shift to social organization through culturally constructed kinship relations provided for stable modes of social organization in hunter-gatherer—and subsequent—societies that transcend what is possible when social organization is dependent on face-to-face interaction.

To account for this transformation from experience-based to relation-based modes of social organization, we need to determine what might be its underlying evolutionary trajectory. Although the details of the evolutionary trajectory are necessarily speculative, we know that it begins with the forms of social organization typified by the OW monkeys discussed in chapter 2 and eventually arrives at the forms of relation-based social organization found among hunter-gatherer groups that we discussed in chapter 3. The trajectory, though, is not simply one of extending trends in non-human primate forms of social organization leading from the OW monkeys to the great apes and the chimpanzees. That trend is characterized, as discussed in chapter 4, by incorporation of increased individualization of behavior through more intensive face-to-face interaction. Increased individualization of behavior substantially increases the complexity of a social unit, and the combination of both increased complexity and individualization appears to reach a limit regarding processes such as biological kin selection and face-to-face interaction for introducing the traits upon which integration of complex social environments depends. The trajectory leading to *Homo sapiens* is characterized, instead, by the introduction of a new mode of social integration based on social relations and does not consist of just further elaboration on prior modes of social integration. It is this transition from one mode of social integration to another that we need to account for. Our goal in this chapter, then, is to identify a plausible trajectory going from experience-based to relation-based forms of social organization that incorporates increased individualization within social organization without reverting to smaller social units.

Forms of Social Behavior and Forms of Social Organization

We will begin by considering the way changes in social organization have related to changes in social behaviors, especially the form of behavior that relates to relation-based social organization. We will distinguish four forms of social behavior between two individuals and their associated forms of social organization. These social behaviors are: (1) asocial, (2) action/reaction, (3) interaction and (4) social interaction between individuals. These four forms of social behavior do not exhaust all possibilities, but we can relate them to major changes in social organization among the ancestral primate species that evolutionarily led to *Homo sapiens*. Of these four forms of behavior, the fourth is uncommon among the non-human primates and only partially incorporated among the chimpanzees. It only fully becomes central to systems of social organization with the appearance of our species, *Homo sapiens.*

These four forms of behavior are implemented in what we will refer to as *dyadic behavior episodes.* By this we mean a sequence of two behaviors in which one individual engages in a behavior and another individual acts in response to the behavior of the initiating individual. The same individual may sometimes be the initiator of a dyadic behavior episode in one context and sometimes the responding individual in another context. When individual A initiates the dyadic behavior episode with behavior b_A (where b_A is a label for the kind of behavior engaged in by individual A, such as cooperative versus self-oriented behavior, not the specific behavior engaged in by individual A), we will refer to the initiating behavior by A as the *prior* behavior by individual A. When individual B is the responding individual in a dyadic behavior episode and does behavior b_B in response to the behavior b_A initiated by individual A, we will call this the *post* behavior by individual B.

We will distinguish among the four forms of social behavior identified above by considering a temporal event consisting of a dyadic behavior episode initiated by individual A with a prior behavior and ended by individual B with a post behavior in response to A's prior behavior. For this dyadic behavior episode, we will consider the way the probability of individual B engaging in a post behavior varies with the kind of prior behavior engaged in by individual A. These probabilities need not be fixed. For behaviors that are subject to learning, the current probability of a behavior by an individual may take into account the past consequences the individual experienced when engaging in that behavior. We can now define the four forms of social behavior using different ways the probabilities for post behaviors relate to the occurrence of a prior behavior.

Asocial Behavior

For *asocial behavior,* the post behavior by B is independent of the prior behavior of another individual, A.[1] This means that individual B does post behavior b_B with a probability that disregards the particular prior behavior, b_A, engaged in by individual A. If all individuals engage in asocial behavior, the result will be a solitary form of social organization, and the distribution pattern of individuals in space will tend to be unpatterned after taking into account constraints such as resource location, physical limitations on the spatial location of individuals, and the distribution of predators. Among the primates, many of the prosimian species have an asocial form of social organization described as solitary but social due to foraging individually but nesting together for sleeping (Sussman 2003). Asocial behavior does not characterize any of the Old World monkeys (*Cercopithecoids*) or the New World monkeys (*Ceboids*). Asocial behavior, also characterized as solitary but social, does occur among the great apes. Each orangutan (*Pongo*), male or female, has its own home range. Home ranges of females overlap extensively, but they avoid encounters (Knott et al. 2008). When there are chance encounters, females interact with behaviors ranging from aggressive to affable (Galdikas 1984). Chance encounters between males are usually agonistic. When mating, a consortship group consisting of the mating pair and her infant or juvenile offspring may form for a few days to several weeks (Cawthon Lang 2005, and references therein).[2] For *Pan troglodytes*, females (at least in East Africa) have been characterized as being asocial in comparison to the highly social behavior of males. East African female chimpanzees are solitary in a manner comparable to that of orangutan females (Wich, Sterck, and Utami 1999).

Action/reaction Behavior

For action/reaction behavior, the probability of a post, or reaction, behavior by individual B may depend on the prior, or action, behavior of another individual A. Accordingly, individual B is reacting to the action of individual A. The probabilities may be fixed when they are genetically encoded and, from an evolutionary perspective, individuals do not yet have the cognitive capacity for updating probabilities through learning. Action/reaction behavior may be asymmetric since an individual can react to the behavior of another individual who is acting asocially. A response behavior can be socially negative; e.g., the reaction, b_B, of B to action, b_A, by individual A, may be to move away or dissociate from A, thereby leading to the spatial dispersal of individuals. Alternatively, a response behavior can be positive; e.g., the reaction of B may be to move towards A, in which case the social structure is pushed towards, minimally, herding or flocking behavior, where each individual acts positively to a few neighboring individuals as discussed in chapter 4. Group boundaries due to a combination of association or dispersal may arise from the probability values for action/reaction behavior.

Interaction Behavior

Interaction behavior depends on having an evolved cognitive system capable of behavior learning. We see this kind of behavior with the OW monkeys. A female OW monkey learns her place in the dominance hierarchy through the responses of other females to her encounters with them.

Interaction behavior differs from action/reaction behavior by the fact that the probability of a prior behavior by individual A is based on learning the probability of a particular post behavior by B based on previous encounters with B. The capacity for learned behaviors can also affect the occurrence of post behaviors when individual B learns to take into account the consequence of doing post behavior b_B in response to the prior behavior b_A by individual A. Like action/reaction behavior, interaction behavior can be asymmetric with individual A acting according to her/his experience with B while individual B may just be reacting in response to behavior b_A by individual A. Learning makes it possible for more elaborated forms of social organization to arise than would otherwise be the case. As we have discussed in chapters 2 and 4, the troop structure of the Old World monkeys (as well as the New World monkeys) and the community structure of the chimpanzees are all based on interaction behaviors.

Social Interaction

Our fourth dyadic behavior elaborates on Talcott Parson's definition of social interaction: "In the case of interactions with social objects a further dimension is added. Part of ego's expectation ... consists in the probable reaction of alter to ego's possible action, a reaction which *comes to be anticipated in advance* and thus to affect ego's own choices" (Parsons 1964: 5, emphasis added). Social interaction differs from interaction because individual A takes into account, before doing prior behavior b_A, what A believes to be the likely response b_B by individual B should individual A do behavior b_A. This contrasts with interaction behavior where A takes into account the likelihood of B doing post behavior b_B independent of A's prior behavior. Interaction and social interaction behaviors also differ regarding the consequences of asymmetric dyadic behavior.

Asymmetric *interaction* behavior by A with action/reaction behavior by B can be advantageous to A since s(he) still takes into account past behavior by B, whereas B does not take into account the past behavior of A. As a result, A uses more information about the behavior of B than B uses about A in responding to the behavior by A. However, asymmetric *social interaction* behavior by A can be disadvantageous to A if A incorrectly assesses B's likely response to a behavior by A. In addition, B may revert to a novel and unexpected behavior in response to A's assessments of B's behavior and thereby engage in a behavior unantici-pated by A. For example, if A assesses that B should act cooperatively and therefore starts to act cooperatively towards B, B may, unexpected-ly, respond with noncooperative behavior as a way to gain benefit from A's cooperative behavior without any immediate cost to B.

Social interaction will be more effective when it is symmetric and leads each agent to engage in the behavior anticipated by the other. If A and B each correctly assess the likely behavior of the other indi-vidual, each will be biasing independently his or her own behavior in the direction anticipated by the other individual, and so any updating of the assessed likelihood of behaviors by A and/or B simply reinforces their current behaviors. With symmetric social interaction, coopera-tive behavior will be reinforced, for example, when each individual acts under the assessment that the other individual will act cooperatively.[3] In this situation, both individuals receive whatever is the payoff for jointly cooperative behavior, thereby reinforcing their respective assessments of the other individual's behavior. However, continuing with symmetric social interaction is not certain, and one or the other of the interact-ing individuals may revert to an unanticipated behavior. Consequently, we need to consider the limitations on conditions for maintaining sym-metric social interaction in conjunction with experience-based forms of social organization.

Limitations of Experience-Based Symmetric Social Interaction

Social interaction as a learned behavior by A depends both on (1) repeat-ed encounter outcomes in which individual A can track the response of individual B when A has engaged in prior behavior b_A in an encoun-ter with individual B and (2) A having a sufficiently evolved, cognitive learning system so as to be able to estimate the likelihood of post behav-ior b_B by individual B when individual A engages in prior behavior b_A. The ancestral, non-human primate taxa on the phylogenetic pathway leading to *Homo sapiens* include increasingly evolved cognitive learning systems, hence we might expect social interaction to have already been introduced within the more cognitively developed non-human pri-mates prior to the evolution of the hominin ancestors to *Homo sapiens*. However, acting against this possibility is the trend towards increased individualization of behavior discussed in chapter 4. The latter trend makes learned behavior more difficult as a basis for interaction of indi-viduals in the same group, thus reducing the likelihood of conditions being present under which learned social interaction would arise even with an evolved cognitive learning system.

As we have seen, conflict between these two trends comes to the fore within the chimpanzees. Their form of social organization has evolved from a prior, cohesive troop form of social organization based on learned interaction into a form of social organization characterized by communities. The internal dynamics of a community are more complex and internally less cohesive than a troop. As discussed in chapter 4 a chimpanzee community is characterized by

- dispersal of females at time of puberty;
- significantly more frequent and longer rates of grooming with more time spent in social dyads by males in comparison to females;
- variation in sociality of females: asocial females in East Africa, more social females in West Africa;
- temporary, fission-fusion subgroups composed primarily of males;
- unstable male dominance hierarchies;
- extensive grooming of adult males, especially upon subgroup reformation;
- high levels of conflict within communities (female-female conflict over access to food and defense of offspring, male-male conflict over dominance rank, and male-female conflict over sexual access); and
- highly aggressive and violent community territorial defense by males that can lead to inter-community killings.

Rather than a cohesive, well-integrated, social system like the troop as a social unit among the Cercopithecoids, a community does not function as a social unit and is characterized by instability of social subunits within the community.

Embedded within this overall pattern is the increased role of learned social interaction in forming social subunits: "Male chimpanzees use grooming to cultivate and reinforce social bonds with others upon whom they rely for coalitionary support" (Muller and Mitani 2005: 306). These dyads are not based on biological kin-relatedness (Mitani, Watts, and Muller 2002). Instead, "Individuals belonging to the same age cohort may be particularly attractive social partners because they grow up together, are generally familiar with each other, and share similar social interests and power throughout their lives" (Mitani, Watts, and Muller 2002: 14); that is, the two males forming the dyad have had sufficient interaction with each other to be able to assess correctly the likely behavior of the other individual. Another way social interaction through learning has been proposed to take place is through sharing meat after killing a prey: "[One] hypothesis proposed to explain meat sharing implicates the use of meat as a political tool.... Male chimpanzees share meat strategically with others in order to curry their favor and support" (Mitani, Watts, and Muller 2002: 18). Learned social interaction underlies patrolling community boundaries: "Males who patrol together also groom and form coalitions with each other frequently" (Muller and Mitani 2005: 308) and patrol "with partners with whom they have strong social bonds and on whom they can rely to take risks" (19).

In brief, the non-human primates present us with a phylogenetic, evolutionary trend of individuals incorporating more precise information about the behavior of other group members while collective, social behavior increases in complexity in response to increased individuation.[4] When symmetric social interaction takes place, each individual acts in the manner anticipated by the other individual, thereby reinforcing coordinated behavior, but asymmetric social interaction behavior can arise through the well-known problem of cheaters. Either party to symmetric social interaction can cheat and revert to a post behavior action/reaction strategy that may be more beneficial, at least in the short run, then social interaction behavior.

Symmetric social interaction based on face-to-face interaction is also costly to learn and must be maintained constantly: "Given the importance of coalitions, *male chimpanzees work hard* to obtain this valuable social service" (Muller and Mitani 2005: 314, emphasis added). Means for maintaining valuable social relations also include reconciliation behavior after conflicts (Kappeler and van Schaik 1992; de Waal 2000). The work required to maintain symmetric social interactions implies that this kind of behavior, when based on face-to-face interaction, does not provide an effective and stable behavioral basis for social groups. The solution to forming large, cohesive groups based on symmetric social interaction that incorporates increased individuation of behavior was only found during the evolution leading to *Homo sapiens* through an evolutionary change from the experience-based social interaction we see in the chimpanzees to a constructed, relation basis for symmetric social interaction.

Cognitive Basis for Constructed Systems
of Symmetric Social Interaction

The evolutionary pathway undertaken by our hominin ancestors leading from experience- to relation-based social interaction builds on five cognitive abilities. These are the ability to: (1) recognize self as a distinct entity, (2) have a "theory of mind," (3) form categorizations based on the concept of a relation between individuals, (4) conceptualize when one relation is reciprocal to another relation, and (5) form new relations through recursive composition of relations. The first of these five abilities is within the cognitive range of the chimpanzees. The evidence for chimpanzees having the second ability is equivocal. It is likely that chimpanzees have the third ability since macaques and vervet monkeys apparently can form categories based on close, biological relations. However, even if chimpanzees form categories based on biological kin relations, the fourth and fifth abilities seem unlikely for them due to the size of working memory that is required for the implementation of these behaviors. In particular, the ability to cognize recursively does not occur with the chimpanzees (Hauser, Chomsky, and Fitch 2002; Corballis 2007), as it is beyond their cognitive capacity due to the limited size of their working memory (Read 2008c); chimpanzees are not able to conceptualize the critical concept of a relation of a relation (discussed in more detail below).

Before I outline a possible evolutionary pathway leading to relation-based social interaction built around these five cognitive abilities, we need a few definitions and conventions for expressing more formally a relation conceptualized as holding between pairs of individuals.

Formal Representation of Dyads and Relations

First, for notation, we will use bold capital letters such as **R** or **S** to name a relation when it does not have a specific name such as the *mother relation* or the *father relation*.

Second, with a relation **R** we can associate the collection, or category, of all dyads from a population of individuals satisfying the criterion (or criteria) for when that relation holds between a pair of individuals. For example, the *mother relation* as it is culturally understood by Americans consists of dyads in which one individual in the dyad is the mother as she is recognized culturally by Americans of the other individual in the dyad. I include the cultural restriction since the criterion for being publicly recognized as mother of someone is culture specific and not universally the same in all societies.

Third, we will symbolically express a dyad composed of a pair of individuals A and B by the notation (A, B). We will use the convention that a dyad (A, B) will be in the category of dyads determined by the relation **R** when the first entry, A, in the dyad is the focal individual to whom the relation **R** will be applied and the second entry, B, is the individual associated with the focal individual by the relation **R**. We also need a convention for the order of the elements in a dyad since a relation such as the mother relation is not symmetric: if B is the mother of A then A is not the mother of B. The first individual in the dyad is the focal individual, and the second person is the individual associated with the focal individual by the relation **R**.

Next I discuss in more detail the five cognitive abilities that are critical for developing the evolutionary pathway leading to relation-based social interaction.

1. Concept of Self

By *concept of self* I mean the cognitive awareness of one's existence, or identity, distinct from awareness of the existence of others. A common experiment used for deciding if an individual has a concept of self is the mirror experiment. A mark of some sort is placed surreptitiously on an individual in a location that cannot normally be seen. The individual then sees her/himself in a mirror. If the individual reacts by touching the location of the mark on him or herself, it is assumed the individual is self-aware. According to this experiment, human infants begin to have self-awareness by around 18 months of age and chimpanzee infants after about 24 months (Bard, Todd, and Bernier 2006). Overall, the experimental evidence for a concept of self is substantial for the chimpanzees and orangutans, though not for monkeys or gorillas (Schilhab 2003). Consequently, we will assume that a concept of self was already present in a primate ancestor common to the chimpanzees and *Homo sapiens*.

2. Theory of Mind

By a *theory of mind* is meant not only that one has awareness of one's own mental states and one's own mental representations as a basis for one's actions, but that one is able to conceptualize that others may have the same awareness of mental states and/or mental representations as a basis for action. Experimental work on human infant development shows that humans can attribute mental states and representations to others as the basis for acting or making decisions by about 4 years of age. The evidence for a theory of mind capability in chimpanzees, though, is equivocal

(Heyes 1998; Povinelli and Vonk 2004). However, there is general agreement that while chimpanzees may not have a theory of mind capacity comparable to humans, they are capable of reasoning about behavior and acting on that reasoning (Call and Tomasello 2008). What is less clear is whether they are able to attribute and reason about mental states in others (Focquaert, Braeckman, and Platek 2008). Given this evidence, we will assume that if a cognitive capacity for theory of mind was not already present in a common ancestor, then it appeared relatively early in hominin evolution.

3. Categorization Based on Relations

By *categorization based on relations* I mean forming categories based on a conceptual relation linking pairs of individuals rather than on properties or attributes of individuals. Categorization based on relations seems to occur to some extent among the non-human primates. One experiment with long-tailed macaques (Dasser 1988a, b) showed that they seem to categorize using a mother relation. In this experiment the macaque subjects were first trained to distinguish between photos of biological mother/offspring pairs and unrelated female/offspring pairs. The subjects were then shown photos of new mother/offspring pairs and unrelated female/offspring pairs from their troop. The subjects consistently distinguished the photos of mother/offspring pairs from the other pairs, indicating that they were distinguishing the relation between two individuals and not the attributes of individuals. Experiments with vervet monkeys and baboons also suggest that they recognize dyads based on the social relationship between pairs of individuals (Cheney and Seyfarth 2007).

For all three of these monkey species, the simplest explanation for the observations is categorization based on the behaviors we refer to as female parenting between a female and her offspring. Female parenting is repeated across female/offspring pairs and would include behaviors such as nursing an offspring, controlling which females may hold an offspring, intensive grooming of offspring, providing emotional and physical support when an offspring has an agonistic encounter with another primate, and so on. These are not behaviors that a female has towards the infant of any female, but are behaviors directed by a female towards her infant, hence the behaviors mark those dyads where the pair of individuals stand in a biological mother/biological offspring relation to each other and so categorization based on those behaviors would be categorization of biological mother/

biological offspring dyads. The alternative, that the categorization is based on something like cognitive ability to track biological relations, can be rejected as it does not account for the way the categorization ability would arise initially.

Whether chimpanzees also categorize based on relations between individuals is, at present, unknown, although they are able to distinguish among pairs of individuals using phenotype matching, a possible precursor to relation categorization. Subject chimpanzees were able, after training, to associate a picture of a novel female chimpanzee with a picture of her biological son or daughter statistically more often than by chance (Vokey et al. 2003). Given this evidence and the fact that several species of monkeys are able to categorize using relations, it is likely that chimpanzees can, too, and hence our last common ancestor with the chimpanzees had the cognitive capacity to categorize based on behaviors consistently associated with close biological relations. For our purposes here, we will assume that categorization of this kind either was already cognitively possible for a last common ancestor with ourselves and the chimpanzees or else it arose early in the evolution of *Homo sapiens*. Since behaviors, such as female parenting, that are a likely basis for initial categorizations provide positive emotional and physical support, we will assume that the initial categorization relations have associated with them the expectation of behaviors we can characterize as positive and supportive.

4. Relation Formation Through Recursive Computation of Relations

The cognitive ability to use a conceptual relation linking pairs of individuals as the basis for categorization is a pre-adaptation for a critical, evolutionary development in the evolution of *Homo sapiens*. It makes possible the formation of, and ability to reason about, new relations from already defined relations through recursive composition of relations rather than through experience. This gives rise to a "functional discontinuity between human and nonhuman minds" since we are the only animals that can "reason about higher-order relations in a structurally systematic and inferentially productive fashion" (Penn, Holyoak, Povinelli 2008: 128). In contrast, categorization based on attributes of objects depends on experience with those objects. New, nonhierarchical, attribute categories cannot be inferred from existing attribute categories alone. Recursive reasoning, though, makes it possible to form new relations and relation categories directly from current, already cognized relations.

We can illustrate the way in which a new relation may be constructed from an already cognized relation through *recursive composition of relations* by considering how a family tree is constructed. Assume we already know that a mother relation assigns to a given person that person's mother. We can now form a new relation, call it *mother's mother*, by defining it recursively from the mother relation. We do this by starting with the focal individual for the family tree. For convenience in reference, we will refer to the focal individual as *ego*. Then we apply the mother relation to ego and trace to ego's mother. Next we take ego's mother as ego, apply the mother relation to this new ego and trace to (ego's mother)'s mother; that is, to ego's (mother's mother). We now define the *mother's mother* relation to be the relation that assigns to ego the female determined in this manner.[5]

Other relations such as mother's father, father's mother, or father's father may be constructed recursively in a similar manner if the father relation has already been cognized. The recursive computation may also be continued further to define the relation, mother's mother's mother, and so on. For each of these new relations, we can form the category of dyads in which the second person in the dyad satisfies the recursive composition relation with the first person in the dyad. For example, from the mother's mother relation we can form the category of dyads such that the second person in the dyad has the relation, mother's mother, to the first person in the dyad. Unlike attribute categorization, we can form new relation categories through recursive composition from already identified relation categories.

Chimpanzees (and other non-human primates) lack the cognitive capacity for recursive reasoning (Spinozzi et al. 1999), let alone composition of relations, due to working memory of size 2 ± 1 (Read 2008c). The recursive composition of even a relation with itself requires a working memory of around 5, as it requires keeping in mind simultaneously: (1) the relation $\mathbf{R}$, (2) the focal object, A, (3) the object B to which A is paired by the relation $\mathbf{R}$, (4) the object B conceptualized as a new focal object for the relation, and (5) another object, C, paired by the relation $\mathbf{R}$ to B. Working memory of size 5 does not appear in our ancestry until around 500,000 BP, namely with our ancestral forms classified as *Homo erectus* (see Figure 5.1).

5. Reciprocal Relations

The last cognitive ability we consider—and one central to the development of cultural kinship systems based on social interaction—is our

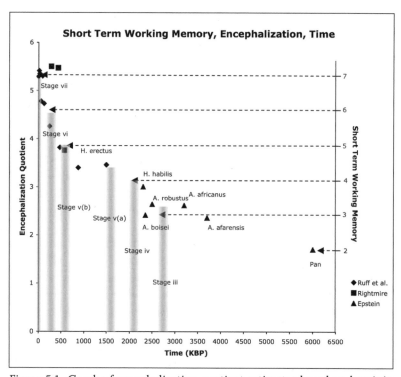

Figure 5.1 Graph of encephalization quotient estimates based on hominin fossils and *Pan*. Hominin fossils have been identified by taxon. Each data point is the mean for the hominin fossils at that time period. Stages iii–vii are stages in the increasing cognitive complexity of the technology used in the manufacture of tools (including tools made by chimpanzees) and vary, for hominins, from Stage iii (conchoidal flaking technology) to Stage vii (recursive technologies) (see Read and van der Leeuw 2008 for details). Height of the "fuzzy" vertical bars is the hominin encephalization quotient corresponding to the date for the appearance of the stone tool stage represented by the fuzzy bar. Right vertical axis assumes a linear rate of change in short term working memory, starting with the working memory of chimpanzees (see Read 2008c). *Data are from Epstein 2002; Rightmire 2004; and Ruff, Trinkhaus, and Holliday 1997. EQ = brain mass/(11.22 x body mass$^{0.76}$) (Martin 1981). Reprinted from Read and van der Leeuw 2008: Figure 2.*

ability to conceptually reverse the relation one individual has to another individual and thereby conceptualize a relation of the second individual to the first individual. In other words, the cognitive ability to recognize the way a relation potentially has a reciprocal relation associated with it. To illustrate, assume the individuals in a group already conceptualize a *mother relation* based on female parenting behavior. Suppose the pair of individuals (*A*, *B*) satisfy the mother relation and so *A* and *B* are conceptualized by "*A* has *B* as mother" when *A* is the focal individual. Suppose attention now shifts to individual *B* and *B* is seen as the focal individual for this pair of individuals. This means that the perspective on this pair of individuals has changed to conceptualizing the way *A* is linked to *B* when *A* and *B* satisfy the mother relation and *B* is linked to *A* by the mother relation. The conceptual link of *A* to *B* would be something like "*B* has *A* as 'a person who has *B* as mother.'" The phrase 'a person who has *B* as mother' is simply what we call the *child relation*. That is, we can consider *child* to be the name for the relation 'a person who has *B* as mother.' With this as the meaning of child, it follows that if (*A*, *B*) satisfies the mother relation, then (*B*, *A*) satisfies the child relation.

In general, the mother relation has as a reciprocal relation either a daughter relation or a son relation when the reciprocal relation takes into account the sex of the initial person for the mother relation. From the viewpoint of behavior, this is equivalent to a female differentiating in her behavior towards a male offspring versus a female offspring. From the OW monkeys and the chimpanzees, although parenting behavior with a new infant does not initially differentiate between male and female offspring, differentiation in behavior does arise as the infant matures. For female philopatric OW monkey species, grooming of male offspring becomes less frequent as they sexually mature and attention is paid primarily to female offspring. For male philopatric *Pan*, among the Pt chimpanzees a biological mother reduces her ties with a female offspring as she reaches sexual maturity and leaves the natal group, but the biological mother maintains a close relationship with her male offspring (Maryanski 1987, and references therein). For the Pp chimpanzees, the connection between biological mother and male offspring extends into adulthood through grooming, providing support in his agonistic encounters and engaging in behaviors that have the effect of enhancing his mating success (Surbeck, Mundry, and Hohmann 2010). For both female and male philopatry, then, the parenting behavior of a female differentiates between male and female offspring as they mature, and so the reciprocal for the mother relation, once it is cognized, may be cognized as a daughter relation and/or a son relation.

We can generalize the cognitive criterion for knowing that the mother and child (or son and daughter) relations are reciprocal relations to other relations. The generalization follows by noting that since the conceptual pathway going from the mother relation to the reciprocal child relation does not depend on the details of the mother relation, a similar conceptual pathway can be formed for any relation **R** to arrive at a reciprocal relation **S**. Generally speaking, then, two relations **R** and **S** are *reciprocal relations* if it is the case that whenever the pair of individuals (A, B) satisfies the **R** relation, then the pair of individuals (B, A) satisfies the **S** relation.[6]

Conceptualizing two relations **R** and **S** as reciprocal relations requires working memory around size 4 since the individuals and both relations must be kept in mind to recognize that if the pair of individuals (A, B) satisfies the **R** relation, then the same pair of individuals, but in the reverse order, satisfies the **S** relation. Even if chimpanzees conceptualize a relation such as the mother relation and independently a child relation, their limited working memory implies that they cannot conceptualize that the child relation is the reciprocal of the mother relation. Consequently, we will assume that the ability to conceptualize reciprocal relations became part of the cognitive repertoire only after working memory expanded during the evolution leading to *Homo sapiens* to around the size associated with early *Homo erectus* (see Figure 5.1).

From Individual Relations to a System of Relations

Next, we consider how a system of relations defined recursively might arise, from an evolutionary viewpoint, and the implications this had for social organization and the boundaries of a social field. I will construct the argument in two parts. In the first part we will see how recursive composition, along with reciprocity of relations, forms a system of relations that is a precursor to the genealogical concept of kin relations in human societies. A critical aspect of the argument involves a shift from relations initially based on patterns of behavior for dyads composed of individuals biologically related in the same manner, such as cooperative, parenting behavior by females directed towards their biological offspring, to relations constructed through recursive composition that are no longer determined by patterns of behavior between biologically related individuals. A relation constructed through recursive composition is not dependent on there being factually a biological connection between a pair of individuals before that pair may be believed to satisfy the relation. This decoupling from biological criteria is a precursor to the formal system of tracing genealogical kinship found in human societies in which the primary genealogical relations of mother and father and their reciprocal relations of son and daughter need not correspond to biological relations. In many societies, for example, a male may be deemed to be the father by non-biological criteria such as being married to the woman recognized by cultural criteria as mother to the child, regardless of his role in providing, or not providing, the fertilizing sperm.

Chapter 5

The second part of the argument establishes the way in which the composition of reciprocal relations leads to symmetric social interaction. As we will see, social interaction is reinforced by the behavior individuals determine to be appropriate according to the system of relations they believe to hold among the individuals in a social unit such as a residence group. The functionality of such a system of relations for a social group depends on that system of relations being part of the conceptual repertoire of each member of the group.

The culturally defined kinship relations that are central to human societies provide little, if any, functionality to an individual in isolation. Instead, the functionality of a system of kin relations derives from the reciprocity of kinship relations, which depends, in turn, on each person in the group who is socially interacting with other individuals having the same kinship system as part of her or his cultural repertoire. For the speaker to know that someone is his or her uncle, cousin or sister-in-law (using the American kinship system as an example) does not, by itself, provide functionality in the form of likely cooperative behavior or mutual assistance among those recognized by the speaker as cultural kin. The functionality arises from those recognized as kin by the speaker reciprocally recognizing the speaker (and each other) as kin. The recognition of each other as kin must occur in a manner mutually understood by all others in the social group. The person recognized by the speaker as uncle, for example, reciprocally recognizes speaker as nephew or niece and the person recognized by speaker as cousin reciprocally recognizes the speaker as cousin in the American kinship system. The persons recognized by the speaker as cultural kin recognize each other, in turn, as cultural kin in a mutually understood manner.

The functionality arising from the behaviors associated with kin relations has the consequence that the boundary of a social group will be based on the boundaries for mutual recognition of who is kin to whom and need not be identical to the boundary of those who interact with each other on a day-to-day basis. The boundary is determined through having a system of kin relations consistent with computation of kin relations based on composition of relations and recognition of reciprocal kin relations. It is no coincidence that hunter-gather groups identify themselves through the boundaries of those who are kin, or can compute that they are kin, to one another, for this is the social field of persons who have a common understanding of how each person should act towards each other person and the kind of behaviors that each person should receive in return. Through reproduction

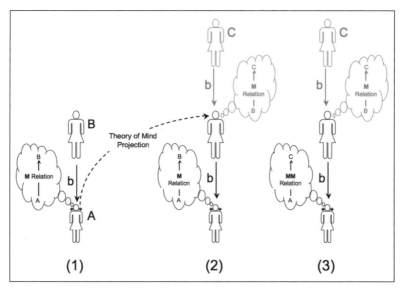

Figure 5.2 (1) *B* engages in parenting behavior, *b*, directed towards her biological daughter, *A*. The **M** relation is derived from categorizing based on female parenting behavior. As indicated by the thought cloud, individual *A* perceives that the **M** relation holds between her and her biological mother. (2) By Theory of Mind projection, *A* believes that there is a female, *C*, for whom *B* recognizes that the **M** relation holds between *B* and *C*. The portions of (2) that are part of the beliefs of *A* are shown in gray. (3) The recursive composition, **MM**, of the **M** relation *A* has to *B* and the **M** relation *A* believes is held by *B* with respect to *C* via Theory of Mind projection connects *A* to *C*.

coupled with enculturation—which occurs through social interaction from birth and is the cultural information transmission process analogous to genomic information transmission process carried out through sexual reproduction—the social group perpetuates itself through time, and the system of kin relations provides a marker for those with whom social interaction is likely to be reciprocated. Those who are one's cultural kin are those who have been encultured into the same cultural system and thus share similar concepts and notions about proper and moral behavior. For a hunter-gatherer group such as the !Kung san, "There is little evidence ... that it is a part of human psychology [for them] to be willing to engage in altruism ... in a social and cultural vacuum. When the *faces and forces of culturally defined institutions* are reintroduced, sharing and giving resume" (Wiessner 2009: 137, emphasis added). Although cultural elements are, of course, located in the minds of individuals, cultural kinship is a group-level and not an individual-level property, and hence provides the basis for organizational change through change in the properties of the cultural kinship system rather than through change at the level of individual traits derived from the cultural kinship system.

Part 1: From Theory of Mind Projection to a Precursor of Genealogical Kinship

Theory of Mind Projection

We will begin by assuming a single relation, the **M** ("mother") relation based on categorization of biological mother/offspring dyads linked through female parenting, is already part of the cognitive repertoire of individuals. (Note that the argument will apply equally to any relation that characterizes dyads among individuals and not just the mother relation.) Assume we have a set of individuals, each having the five cognitive abilities discussed above as part of his or her cognitive repertoire. In Figure 5.2(1), female A, the biological daughter of female B, conceptualizes the relation between herself and her biological mother as satisfying the **M** relation based on, say, parenting behavior b, so A perceives B as "my mother." Because of the Theory of Mind, A also perceives the situation as if she were in the position of her mother. By assuming her mother's perspective, she believes that there is a woman C her mother would refer to as "my mother," based on behavior b. Thus the (B, C) dyad is believed by A to satisfy the **M** relation from her mother's perspective (see Figure 5.2[2]). This is a *belief* from the perspective of A since A projects onto her mother A's belief that her mother also conceptualizes the **M** relation. Correspondingly, the thought cloud in Figure 5.2(2) is dashed and in gray for female B to indicate that this is the conceptualization A believes is held by her mother. This may, or may not, correspond to what is actually conceptualized by her mother.

Theory of Mind and Recursive Composition of Relations

When individual A conceptualizes and projects the **M** relation it follows that individual A can construct the **MM** relation through recursive composition of the **M** relation A has with respect to B and the **M** relation A believes that B has with C. The **MM** relation enables individual A to believe that she and individual C are conceptually connected (see Figure 5.2[3]). The **MM** relation constructed through recursive composition differs in a crucial way, though, from the **M** relation. The **MM** relation is constructed from the **M** relation and not from a categorization based on behaviors between actual biological grandmother/biological granddaughter dyads. Although the **M** relation arises initially from behaviors between biologically related individuals, recursive composition of **M** leads to the construction of a new relation, **MM**, without depending upon a prior categorization of dyads based on behavior between biologically related individuals. Instead, the categorization of dyads is now a *consequence* of the new relation formed through recursive composition. The new relation includes all those dyads (A, C) where, by virtue of the Theory of Mind, individual A projects onto individual B (who satisfies the **M** relation from A's perspective) the relation, **M**, and through this projection individual A *believes* that individual C is the target of the **M** relation for individual B. Hence once constructed, the **MM** relation gives rise to a category of dyads for A that are A's perception of dyads that satisfy the **MM** relation.

It is A's belief, by way of Theory of Mind, that is critical in this process, not the actual behavior between C and B. In other words, a recursive composition leads to a new relation from existing relations. Since a recursive composition can be repeated, it can create a system of new relations constructed from an initial relation or relations. One of the consequences of constructing a new relation such as **MM** using recursive composition is that the newly constructed relation no longer depends on the biological facts of who is related genetically to whom, but only on beliefs held by individuals about what allegedly are the biological facts. The system of relations **M, MM, MMM, …** constructed through recursive composition leads to categorization of dyads constructed not from behaviors between individuals but from the relations generated through recursive composition. The categorization no longer depends, as does the **M** relation, on categorization based on behaviors between biologically connected individuals. *Recursive composition of relations leads to decoupling of constructed relations from a biological substrate when categorizing is based on the constructed relations.*

Theory of Mind and Reciprocal Relations

Next we consider the reciprocal daughter relation **D** (or son relation **S**) for the mother relation **M** between A and B. From the perspective of B, A will be the target of the **D** relation. Thus individual B perceives the reciprocal **D** relation with A (A is "my daughter" from B's perspective) (see Figure 5.3[1]). Suppose B projects the **D** relation onto individual C (see Figure 5.3[2]). Due to this projection, B *believes* that individual C perceives the daughter relation **D** with herself as the target of the **D** relation from C's perspective. Consequently, B will believe that B and C are conceptually linked to each other through the **D** relation and reciprocally through the **M** relation.

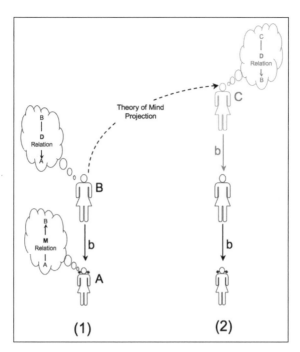

Figure 5.3 (1) *A* perceives the mother relation **M** between *A* and *B* based on female parenting behavior *b* and *B* perceives the reciprocal daughter relation **D** between *B* and *A*. (2) In a Theory of Mind projection, *B* believes that *C* perceives the daughter relation **D** between *C* and herself and will engage in behavior *b* (shown with a gray arrow).

We now have precursor conditions for social interaction from *B*'s perspective; namely, *B* believes *C* perceives the reciprocal **D** relation with *B* and can act accordingly, hence *B* may act towards *C* in anticipation of *C*'s behavior towards *B*. Although illustrated with the **M** and **D** relation, the same pattern will arise for individual *B* with any relation **R** that *B* has with *C* when there is a corresponding reciprocal relation that *B* may have with respect to an individual *A*.

Precursor to a System of Genealogical Relations

Recursive composition of relations applied to the **M** and **D** (or **S**) relations leads to a system of generated reciprocal relations. The relations **MM, MMM,** ... have **DD, DDD,** ... (or **SS, SSS,** ...., **SD, DS,** ...) as their reciprocal relations. Recursive composition of **M** and **D** (or **S**) relations creates the **MD** (or **MS**) relation. This recursive composition has as its reciprocal relation the recursive composition **DM** (or

SM) relation. Each relation generated through recursive composition will have a reciprocal relation. Thus a repeated recursive composition acting on the **M** relation and its reciprocal **D** or **S** relation leads to a system of relations closed under composition and reciprocity of relations. Closure under reciprocity is central to cultural kinship systems: if individual *A* is cultural kin to individual *B*, then individual *B* is cultural kin to individual *A*. This makes it possible to form a social unit consisting of individuals who are cultural kin to each other. Thus, the system of relations generated through recursive composition satisfies a fundamental property of kinship systems, namely that it is a system of mutual relations.

It is a precursor to kinship relations determined through genealogical connections, since the system of relations based on the **M** relation and its reciprocal **D** and **S** relations is, on the one hand, more extensive than the formal system of tracing genealogical connections we find in human societies. On the other hand, it does not have a father relation analogous to the mother relation in which the target individual is a male rather than a female.

The greater extensiveness of the system constructed through unconstrained reciprocal compositions is due to recursive compositions potentially including relation compositions such as *mother's child's father*, whereas in kinship systems determining if two individuals are genealogically connected is usually limited to upward genealogically tracing using the parent relation from each of the two individuals to see if they have an ancestor in common. (In this tracing process the parent and/or child relations may also be sex marked.) The more extensive system of relations constructed through all possible recursive compositions can be formally transformed into the more limited structure of genealogical relations by restricting recursive compositions to either compositions using the **M** relation, compositions using the **D** or **S** relations, or compositions using the **M** relation followed by compositions using the **D** or **S** relations. The formal statement can be expressed as a "rule-of-thumb": tracing is either upwards, downwards, or upwards and then downwards.

Change in Mating Patterns Leading to a Father Categorization

The possibility of a biological father relation analogous to a biological mother relation based on female parenting is not likely to have been part of the cognitive repertoire of a common ancestor to *Pan* and *Homo sapiens*. Male parenting, when it occurs in monkeys and apes, appears to relate to increased mating opportunities rather than parental investment (Smuts and Gubernick 1992). Consequently, the concept of a biological father relation must have arisen during a later time in the evolution of *Homo sapiens*. We can identify the conditions under which this might have occurred by considering changes that took place in food procurement between males and females as part of the evolutionary development of *Homo sapiens*. These changes led to selection for temporary to permanent male-female bonding, including emotional ties, as part of the overall pattern of environmental and social adaptation that developed among the ancestors of our species.

I will sketch out, briefly, those aspects of the adaptation that relate to behaviors leading to the conceptualization of a father relation. We start with bipedalism as an early hominin trait that marks the beginning of the distinctive mode of adaptation that eventually led to *Homo sapiens*. Bipedalism is likely to have begun as the primary mode of locomotion 6 or 7 million years ago (Harcourt and Aiello 2004, and references therein) due to its greater efficiency for traversing more open environments in comparison to the kind of quadrapedalism or occasional bipedalism used by chimpanzees (Sockol, Raichlen, and Pontzer 2007). Evidence for a bipedal posture comparable to that of modern humans is more recent, though, and dates back to 3.6 million years ago as shown by the features of the footprints at the hominin site of Laetoli in Tanzania (Raichlen et al. 2010).

For females, the shift to bipedalism constrained the size of the pelvic birth canal (Lovejoy 2005). Subsequently, selection beginning around 2 million years ago for a relatively larger, hence encephalized, brain size in comparison to body size (Klein 1999), in combination with the constrained birth canal, led to what is called secondary altricial offspring; that is, offspring requiring extensive care and parenting after birth due to restricted physical and neurological development at time of birth. Altriciality in *Homo* appears to have resulted from selection for an increase in adult brain size without a simultaneous increase in gestation length. Gestation time did not increase because an extended gestation time would have produced a fetus too large to pass through the restricted size of the birth canal. As a result, offspring began to be born at a relatively less developed stage than would be expected in comparison to the great apes. Human gestation would need an additional six months for the newborn to have physical maturity at birth matching that of a great ape newborn (Passingham 1975).

Secondary altriciality reduced the mobility of females for an extended time after giving birth (Falk 2004). The restriction on the mobility of females was further accentuated with loss of body hair for thermoregulation (Wheeler 1992), an event dating back at least 1.2 million years (Rogers, Iltis, and Wooding 2004). Loss of body hair meant the offspring could not cling to the biological mother as she moved through the environment. That, coupled with increasingly altricial offspring, required females to carry or otherwise hold on to offspring when walking or running, which further restricted their mobility.

Over approximately this same time period, the hominin diet was changing from a chimpanzee-like frugivorous diet with occasional small hunted animals to a more faunivorous diet that increasingly included parts of carcasses scavenged from carnivore kills using flake and pebble-chopper stone implements. Both passive scavenging (parts of a carcass removed from abandoned kills) and confrontational scavenging (predators driven from a kill) may have been used to obtain parts of carcasses (Potts 2003). The scavenged parts were most likely transported to locations where they could be processed in safety for meat scraps and the bones cracked open for their marrow content (Blumenschine and Pobiner 2007). This increase in meat protein and high-caloric bone marrow in the diet made possible a decrease in the energy requirements for maintaining a large gut, allowing a reduction in its size in comparison to the gut size in non-human primates. The reduction in gut size made available metabolic energy that could be used for the energy requirements of a larger brain (Aiello and Wheeler 1995; Snodgrass, Leonard, and Robertson 2009).

Even with passive scavenging, predators would still be a threat, making it likely that scavenging would "be limited to older subadult and prime adult male hominins" (Blumenschine and Pobiner 2007: 183). The reduced mobility of females with altricial offspring implies that they would have had limited or even no involvement in scavenging. Consequently, meat protein and marrow calories obtained through scavenging could become a regular part of, and not just an occasional addition to, the female diet only if there were a modification in male-female mating relationships that led to consistent sharing of the scavenged resources obtained by a male with one, or possibly a few, females. A precursor, if not a preadaptation, to this shift in mating patterns can be seen in the positive association between copulation rates and sharing of meat by chimpanzee males with females in the Taï Forest community (Gomes and Boesch 2009). However, Pt chimpanzees in other communities do not have the same pattern (references in Gomes and Boesch 2009), and for the Pp bonobos, it is the females that both control and share prey (Hohmann 2009). Consequently, sharing of meat, by itself, does not seem to be sufficient to account for the introduction of male-female bonding that eventually became the basis for regular resource sharing among the hominins.

The initial modification of mating behavior in the direction of male-female bonding may have been reinforced by changes in female sex characteristics that also occurred during this time period. Regardless of the initial, proximal reason for the specific changes, a complex of changes, including concealed ovulation, permanent breasts, female pattern for fat deposition, and continual sexual receptivity and orgasm, all appear to relate to "elicitation of long term male attention" (Sazalay and Costello 1991: 449), which may have been expressed through selection for emotional attachment of a male to a female and vice-versa. Overall, the outcome was likely a shift to mating patterns that included at least temporary pair-bonding, possibly through emotional attachment, between a male and a female. Pair-bonding of this sort would reinforce a behavior of sharing by a male of scavenged and hunted small animals food resources as part of his means for continued sexual access to a female with whom he had emotional attachment. This would also have had the consequence of at least indirectly provisioning his offspring through his relationship to her, if not directly engaging in male parenting through providing food resources for his offspring as is suggested by research on rodents that demonstrates a common neuroanatomical and neuroendocrine basis for pair-bonding, monogamy and paternal parenting (Fernandez-Duque, Valeggia, and Mendoza 2009). As long as temporary pair-bonding increased the likelihood that her offspring were also his offspring beyond a chance level, sharing of scavenged meat and hunted small animals would increase the reproductive fitness of each of the pair-bonded partners.

With pair-bonding like this in place, conditions making it possible to conceptualize a father relation would have been introduced, yet under conditions that differ from those for conceptualizing a mother relation. The father relation could just involve categorization based on positive, supportive behavior directed by a male to the female who is the target of the mother relation, rather than parenting behavior directed by the male towards the focal offspring for the mother relation. A father relation conceptualized in this manner is consistent as a precursor for the way genealogical father in many human societies is based on a male's relation to the genealogical mother and not to his putative offspring.

Part 2: From Theory of Mind Projection to Stable Symmetric Social Interaction

The second part of the argument focuses on the implications of reciprocity of relations for stable, symmetric social interaction. Reciprocity of relations implies that the speaker will be both the focal person for, and target of, either a relation and its reciprocal relation or vice-versa. The same will be true of the projection of a relation and its reciprocal relation through the Theory of Mind. Hence the speaker will anticipate that the target person shall exhibit the kind of behavior associated with the relation (or its reciprocal relation) in response to behavior directed towards the target person. By itself this is not sufficient for stable, symmetric social interaction, since the target person does not act in accord with the speaker's beliefs about the behavior of the target person, but acts in accordance with the facts of the situation. The symmetry arises, I will argue, when the target person also conceptualizes the same relation and reciprocal relation as a speaker, and hence acts in accordance with her/his anticipation of the other person's behavior based on a Theory of Mind projection by the target person. Underlying this scenario is the functionality associated with a projected relation.

Functionality of a Projected Relation: Social Interaction

The importance of perceiving a relation **R** lies not in the relation, per se, but in behaviors that are associated with the relation and thereby lead to social interaction and not just interaction. A behavior such as altruism introduced through biological kin selection is not social interaction when there is no anticipation (implicit or explicit) by the actor that the behavior will be reciprocated in some manner. In contrast, a behavior based on a culturally constructed kinship relation satisfies the conditions for social interaction since the conceptual system that structures cultural kinship (namely a kinship terminology) is part of a system of reciprocal relations, including anticipated reciprocal behavior. We can see this by considering the situation where A recognizes B as a cultural kin; that is, A has a kin term that can be properly used to refer to B, and A knows that B shares with A the same kinship terminology system. Under these conditions, A also knows that B has a reciprocal kin term for A. Consequently, A believes that B recognizes A as a cultural kin and expects B to act accordingly. Thus A can assess the expected reciprocal behavior of B by knowing that the culturally constructed kinship system through which A knows that B is a cultural kin of A is also cultural knowledge shared by B.

When a behavior (or kind of behavior) is viewed as being part of, or associated with, a relation **R** linking a pair of individuals, the behavior will be part of a Theory of Mind projection that one individual forms vis-à-vis another individual. More precisely, suppose that individual *B* has an **R** relation with individual *C* and a reciprocal **S** relation with individual *A*, where the biological relations among *A*, *B* and *C* may be unknown to them (see Figure 5.4[1]). Suppose that individual *B* directs behavior *b* (or the kind of behavior represented by *b*) towards individual *A* when that individual is a target of the **S** relation conceptualized by *B* (see Figure 5.4[1]). For example, for English speakers **S** might be the cousin relation, and *B* engages in positive, supportive behaviors with those who are cousins of *B*. More generally, according to Fortes's kinship Axiom of Amity, *B* may engage in positive, supportive behavior with anyone known by *B* to be her or his cultural kin (such as individual *A*). By the Theory of Mind, when individual *B* projects the **R** relation and the reciprocal **S** relation to individual *C*, then *B* will assess that *B* will be the target individual for the reciprocal **S** relation from *C*'s perspective. In addition, B will assess that the behavior *b* (or the kind of behavior represented by *b*) associated with the relation **S** will be engaged in by *C* and directed towards *B* (see thought clouds in Figure 5.4[2]). Via the Theory of Mind projection, individual *B* believes that individual *C* will engage in the behavior *b* (or in *b*-like behavior) towards *B*, since individual *B* is a target of the **S** relation that *B* believes to be a relation concept held by *C*. Now if individual *B* believes that individual *C* will engage in the behavior *b* (or *b*-like behavior) with respect to *B*, then individual *B* may engage in the behavior *b* directed towards *C* in *B*'s belief that individual *C* will reciprocate with behavior *b* (or *b*-like behavior) directed towards *B* (see Figure 5.4[3]). *We now have a basis for social interaction: one individual acts towards another individual under the belief that the other individual will act in a reciprocal manner. Further, and critically, the basis for social interaction is decoupled from any requirement of biological linkages among the individuals in question.*

Theory of Mind and Reciprocal Relations as a Basis for Symmetric Social Interaction

While the projection of a behavior linked to a relation may lead to the belief that this (or a comparable behavior) will be engaged in by the other individual, the reciprocal behavior need not occur. One reason for the occurrence of the behavior by the other individual would be that the other individual has also constructed a complementary belief system

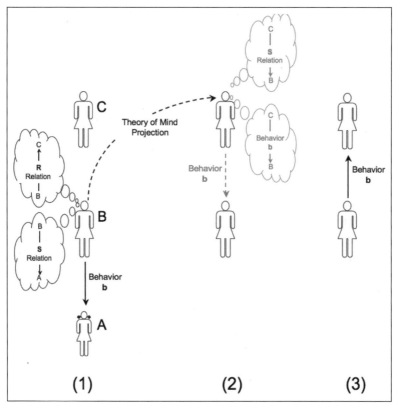

Figure 5.4 (1) Individual *B* has the **R** relation with individual *C* and the reciprocal **S** relation with individual *A*. Individual *B* associates behavior *b* with the **S** relation. (2) By Theory of Mind projection, *B* believes that *C* perceives the **S** relation to *B* and will engage in behavior *b* towards *B*. (3) Based on B's assessment that *C* will enage in behavior *b*, *B* engages in behavior *b* towards *C*, thus individual *B* has engaged in social interaction with *C*.

about the behavior of the initiating individual and for this reason reciprocates with a behavior directed towards the initiating individual. However, cheating, including the situation where the reciprocal behavior is not initiated despite the individual having the complementary belief system, is always possible, and if *B* acts towards *C* solely because of B's *belief* that *C* will reciprocate, then *B* has also initiated conditions that favor cheating by *C*.

Symmetric social interaction can be initiated, however, when each of *A* and *B* recognizes that he or she is conceptually linked to the other through relations with associated behaviors.[7] Under these conditions, each may engage in social interaction by initiating a behavior towards the other, taking into account the expected behavior by the other individual according to the relation that holds between them and its associated behavior. If not, they each lack a basis for expecting reciprocal behavior. In small, kin-based societies, there typically cannot be social interaction between individuals A and B, as discussed previously, without A and B first establishing that they are cultural kin—which means that B is already in the domain of A's cultural kin, A is in the domain of B's cultural kin, and both know that this is true of the other person. Under these conditions each person has expectations about the behavior of the other person based on knowing that the other person is his or her cultural kin. In contrast, with an encounter between two individuals who are strangers to each other, and hence are without a cultural kin relation between them, there is no basis for assessing what would be a likely behavior by the other individual.

In order for there to be a pattern of symmetric social interaction, the individuals must engage in reciprocal behaviors. If each person believes that the other will reciprocate the behavior in question, then, as we will now see, the basis for continued symmetric social interaction arises in a simple way. In order for each of individuals B and C to have the belief that the other individual will reciprocate with behavior b, it suffices for individual C to associate the behavior b with the relation **R** along with individual B associating the same kind of behavior with relation **S**, where **R** and **S** are reciprocal relations (see Figure 5.5[1]). Under these conditions, individual B, through the Theory of Mind projection, believes that individual C will reciprocate with behavior b. Independently, individual C believes, through the Theory of Mind projection, that individual B will reciprocate with behavior b (see Figure 5.5[2]). When each individual engages in behavior b directed towards the other individual according to their respective beliefs, the consequence is that his or her beliefs are reinforced through the *actual* behavior of the other individual. Individual B, for example, believes that C will engage in behavior b directed towards herself or himself, and in fact individual C engages in behavior b directed towards B because C believes B will direct behavior b towards herself or himself, and vice versa (see Figure 5.5[3]). Thus, each of individuals B and C is motivated to direct behavior b towards the other individual as an instance of social interaction, and in so doing, each reinforces and confirms the beliefs that each individual has about the behavior of the other individual.

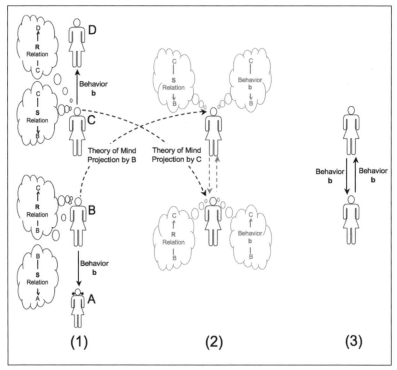

Figure 5.5 (1) Individual *B* has individual *C* as the target for the **R** relation. *B* associates behavior *b* with the reciprocal **S** relation. Individual *C* has individual *B* as the target of the reciprocal **S** relation. *C* has individual *D* as the target of the reciprocal of the **S** relation for which *B* is the target from *C*'s perspective. (2) Individual *B* projects, by Theory of Mind, the **S** relation to individual *C* and believes that *C* will engage in behavior *b* towards *B* (gray dashed downward arrow). Similarly, individual *C* projects, by Theory of Mind, the **R** relation to individual *B* and believes that *B* will engage in behavior *b* towards *C* (gray dashed upward arrow). (3) Each of *B* and *C* acts on her respective beliefs. *B* engages in social interaction behavior *b* directed towards *C* in *B*'s belief that *C* will engage in behavior *b* directed towards *B*, and similarly for individual *C*.

Coordination Through Enculturation and an Expanded Social Field

A critical implication of the argument being made here for stable social interaction lies in the requirement that both the speaker and the target person have the same conceptual system of relations and reciprocal relations. If they do, then the social system will consist of individuals who each have the same conceptual system as part of his or her cultural repertoire. A social boundary formed through a shared cultural repertoire provides the basis for coordination in behaviors necessary for symmetric social interaction. For symmetric social interaction to take place, it is not sufficient that individual A includes individual B within A's cohort of cultural kin. Individual A must also be within B's cohort of cultural kin. Coordination will occur when the culturally constructed kinship system (as well as other cultural idea systems) is part of the cultural repertoire transmitted to each individual through enculturation.

Coordination Through Kinship Computational Systems

Under these conditions, knowing that someone else is one's cultural kin becomes a marker of the fact that the target person has been enculturated with the same cultural repertoire. Conversely, when a person and target person have both been enculturated into the same cultural repertoire, then that person and the target person will both have the same culturally constructed kinship system as part of each person's respective cultural repertoire. Just as restricting reproduction to members of a biological species ensures that all members of the species have comparable biological information systems (though differing from one individual to another in the details of a particular genetic information system), a social group bounded by transmission of a cultural kinship system ensures that all members of the social group have essentially the same cultural information systems (though differing from one individual to the other in the details of a particular cultural information system). All members of *Homo sapiens* have, for example, genetic information systems for a heart, a circulatory system and a breathing system as part of their genome received through reproduction, although the specific properties of the circulatory system or breathing system are variable from one individual to another. Similarly, all members of a hunter-gather group who self-identify themselves as "we the real people" according to a cultural kinship relation as the criterion for being a "real person" also have in common similar cultural idea systems about marriage, ownership of resources, and so on, although possibly differing in detail regarding how these cultural idea systems are implemented.

For the functionality of the reciprocal behaviors associated with symmetric social interaction to be realized, it suffices, then, for the individuals in the population to recognize, in a comparable manner, the same relations and their associated behaviors. When this happens, a relation connecting pairs of individuals becomes an identifier for dyads of individuals who will reciprocate behaviors associated with the relation. Agreement between actor and recipient with respect to engaging in the behavior will occur when the actor and the recipient both associate the behavior, or kind of behavior, with the relation **R** and its reciprocal relation **S**. Consequently, the likelihood of actually realizing the functional benefit that can accrue from the reciprocal occurrence of a behavior depends on the degree of coordination and agreement among individuals regarding relations and their associated behaviors. This coordination is a precursor to the institutionalized social action and role systems prominent in human societies (Nadel 1957), as these "are clothed in cultural meaning systems so that institutions cannot be properly represented without ... reference to shared meanings" (Fararo 1997: 76).

Social Boundary for an Expanded Social Field

A solution to the coordination that underlies stable and symmetric social interaction arose, then, in the evolution of social systems leading to human societies with the transition to cultural kin relations transmitted through enculturation. This development had the effect of expanding the social field beyond those interacting on a face-to-face basis to those who recognized each other as cultural kin even when they were in different social units. The systems of cultural kin relations we find in small-scale human societies today are expressed through the structure of kinship terminology systems as discussed in chapter 3. The importance of expressing kinship relations through a kinship terminology lies in the way this makes kin relations explicit through: (1) forming a system of reciprocal kin relations, (2) having a structure in the form of a generative system, which facilitates faithful transmission of the conceptual system of kin relations across and within generations, and (3) a computational system through which kin relations may be calculated in a straightforward and mutually understood manner.

Because of these properties, when individuals A and B share the same kin term computational system, it follows that once A knows or determines that A has a kin term relation to B, A also knows that B has a kin term relation to A, and in addition A knows that the relation B has to A is the reciprocal (in A's computational system) of the relation that A has to B. To continue with an earlier example, when an English speaker, A, refers to someone else, B, as *uncle*, the speaker not only knows that (s)he is *niece* or *nephew* from that person's perspective, but A also knows that from B's perspective, the relation A has to B is the reciprocal (in B's computational system) of the relation B has to A. The speaker also knows that the other person recognizes that the speaker is *niece* or *nephew* to him, and so he must be *uncle* with respect to the speaker. In other words, *A* knows that both *A* and *B* recognize that they are reciprocally related through the cultural kinship system. The same applies equally to *B* so long as—and this is critical—*B* recognizes the cultural kin relation between *B* and *A* because of *B* also having the same kinship terminology system as part of *B*'s cultural repertoire.

Cultural kinship relations thus become a means for identifying those individuals who have the same cultural repertoire. For individuals who share the same kinship terminology system, the domain of kin as constructed by one person through the kinship terminology can be translated into the way that domain of kin will be constructed by another person who has the same kinship terminology as part of his or her cultural repertoire. Consequently, once *A* knows that *B* is in the domain of individuals with whom *A* has a cultural kin relation and vice-versa, then the belief that *A* constructs regarding *B* (namely that *B* is a person who will reciprocate behaviors directed from *A* to *B* by virtue of the kin relation between *B* and *A*, from *B*'s perspective) will, in fact, be corroborated by the behavior of *B* by virtue of *B* forming a reciprocal belief about *A* from *B*'s perspective (see Figure 5.5[3]). Hence, the conditions necessary for the reinforcement of symmetric social interaction are satisfied when each of *A* and *B* have the same conceptual system of kin relations as part of their respective cultural repertoires.

Predicted Modal Size of the Expanded Social Field

The set of persons who mutually recognize each other as cultural kin (or who can compute that they are mutually cultural kin) form a bounded social system in which symmetric social interaction does not depend on extensive prior face-to-face interaction for knowing whether behaviors will likely be reciprocated and in what manner. That hunter-gatherer groups self-define the boundary of their social field through those who mutually recognize each other as kin (or can compute that they are kin) is not just an interesting observation, but critical, as we have seen, to symmetric social interaction serving as the basis for the social coherence of a group. It underlies the kinds of human social systems that evolved as part of the transition from face-to-face to relation-based systems of social organization.

The boundary of the social group defined in this manner is determined by the maximum size of a group within which individuals either know or can compute their kinship relations to other members of the group. This implies that the social boundary for a hunter-gatherer group will be structured by the boundary for computable kin relations and is not simply a consequence of environmental and ecological conditions. We can test this claim empirically through data on the size of hunter-gatherer groups and whether the group size varies in a patterned way with environmental and ecological conditions or according to the structural basis for a group boundary. If the group size is determined by the size of the social field within which mutual kin relations are known or may be computed, then we can predict the modal size of the social field for hunter-gather groups by a simple argument starting with the modal size of a residence group in the following manner.

Population Size: Empirically Based Modal Value

At first glance we might expect the catchment area (the region from which resources are obtained) to reflect local ecological conditions, in which case the population size of the hunter-gatherer group should not be constant across different hunter-gatherer groups but would vary according to ecological conditions that range from desert to tropical to temperate to arctic conditions. However, Joseph Birdsell (1953) argued to the contrary—using data on hunter-gatherer societies in Australia—that the population size of hunter-gatherer societies has a modal value of about 500 individuals independent of ecological and environmental conditions. This implies that the population size is constrained by the internal, organizational dynamics of hunter-gatherer societies and not by external environmental and ecological conditions.

Birdsell did not identify what the organizational constraints would be. Nonetheless, subsequent researchers have taken the modal value as dogma about hunter-gatherer societies (Kelley 1995) with one notable exception. Lewis Binford (2001) challenged Birdsell's results on empirical grounds: "All of the patterns generated by my analysis strongly contradict Birdsell's conclusions" (223). His results seem to definitively refute Birdsell's argument except for the fact that the data set used by Binford is heterogenous regarding the structure and organization of hunter-gatherer societies. Binford culled data on $n = 339$ societies identified as hunter-gatherers in the literature, along with whatever estimates of population size and area were provided by the author or researcher. All hunter-gatherer societies were treated coequally by him regardless of whether the hunter-gatherer society was "simple" (social organization is kinship based without political offices or institutions) or "complex" (social organization includes political offices and/or institutions). As a result, the range of population sizes in his data set covers several orders of magnitude, ranging from p = 23 (Deep Springs Paiute in Owens Valley, California) to p = 14,582 (Shuswap in British Columbia).

Another, but less problematic, limitation has to with the fact that societies are included in the data set regardless of the extent to which they have been affected by neighboring agricultural groups and/or by the intrusion of outside groups (especially westerners) into indigenous hunter-gatherer lands. Assuming that inaccuracies in data reported by authors and distortions in population sizes caused by contact with non-hunter-gatherer societies are distributed roughly randomly throughout the societies in the data set, these errors mainly introduce noise that reduces the likelihood of finding meaningful patterning. Conversely, when statistically significant patterning is found despite the noise in the data set, there is good reason to pursue the patterning as potentially meaningful.

Data Heterogeneity

The heterogeneity in the data set due to the inclusion of both simple and complex hunter-gatherer societies is more problematic than limitations on the quality of the data since the consequences of ecological conditions and social organization on population size are very different for a complex hunter-gatherer society such as, for example, the Kwakiutl (p = 14,500) than for a simple hunter-gatherer society such as the !Kung san (p = 726) that lacks any political positions or political institutions. As a first approximation, we would expect simple hunter-gatherer societies to consistently have smaller population sizes than complex hunter-gatherer societies.

We can determine whether there is a systematic heterogeneity effect on population size through examining a histogram for the frequency distribution of hunter-gatherer population sizes. A primary signature of a systematic heterogeneity effect is bi- (or multi-) modality in the histogram. Multi-modality indicates that there are different dynamics occurring among hunter-gather societies, and so we would need to divide the data set according to the modes to remove heterogeneity in the data set (Read 2005). Then we would need to determine whether the modes correspond to structural differences such as simple versus complex hunter-gatherer groups.

The complete data set consists of n = 339 hunter-gatherer societies of which n = 54 are considered by Binford to be based on questionable data, and thus have been excluded here. The population size histogram for the remaining n = 285 hunter-gatherer societies is shown in Figure 5.6. There is unmistakeable bimodality dividing hunter-gatherer societies into those with population size p < 775 (see inset, Figure 5.6, anti-mode at 775) and all other hunter-gatherer societies. (There is possibly another anti-mode at p = 4400, but we will not pursue this further as we are concerned only with the first subset of hunter-gatherer societies.) The left group of population sizes has mode p = 500 (which agrees with Birdsell's estimate), mean $\bar{x}$ = 363, and median m = 385. Both the median and mean are displaced downwards by societies with reported population sizes that are too low (e.g., p < 100) for a hunter-gather society to be viable over time, hence the mode at p = 500 is the best estimate of central tendency for this data set.

The data points corresponding to the leftmost mode are simple hunter-gatherer societies found in all ecological and environmental conditions. These data support the argument that simple hunter-gatherer societies have a population size determined by organizational rather than ecological constraints. This being the case, we need to consider what those constraints would be.

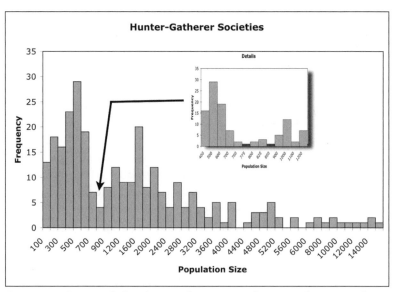

Figure 5.6 Bar graph for the population size of n = 285 hunter-gatherer societies. *Data are from Binford 2001: Table 5.01.*

Chapter 5

Kinship Computation Constraint on Population Size

Hunter-gatherer societies consist typically of persons distributed among a series of residence groups, with each group living in a recognized locality and having both general rights common to all society members and more specific rights deriving from membership in a residence group, such as the right to procure resources in the spatial region associated with that residence group. For simple hunter-gatherer societies, both societal and residence group memberships are framed through kin relations. The hunter-gatherer group that we refer to as the !Kung San, for example, self-identify themselves, as we have already noted, by the expression *ju/'hoansi* (or *ju'/hoasi*, among other phonetic spellings), meaning "the real people" (Lee 1993), where being a real person is to be a kinsman. This kind of self-identification is common for hunter-gatherer societies. The prerequisite for social relations between two individuals in such a society is that they first establish or already know that they are kin to one another. Hence the size of the society is bounded by the size of the set of persons who can determine that they are mutually kin.

Kin knowledge for an offspring begins with the offspring understanding himself or herself as a social self (zero order kin). As the offspring develops, he or she learns about other individuals who are his or her first order kin through interaction with them. This learning process leads eventually to competence in using the kinship idea system. The latter enables two persons, A and B, who do not yet know if they have a kin relation to each other, to determine whether they have a kin relationship by each identifying him or herself as having a first order kin relation to a third person C. Once this third person is identified, they may compute their kin relation to each another as discussed in chapter 1 from knowing their respective first order kin relations to C. We will refer to a kin determined in this manner as a second order kin. A third order kin would be based on a computation in which person A identifies person A' as a kin, person B identifies person B' as a kin (where A' and B' do not yet know if they are kin to each other), and then A' and B' identify themselves as second order kin, thereby enabling persons A and B to compute the kin relation they have to each other. Clearly, third (or higher) order kin term computations will be rare. Henceforth we will assume all kin relations are at most second order.

Population Size: Predicted Modal Value

We will now predict the total population size for a society composed of mutually recognizable kin divided among several residence groups.

We will assume a life-cycle in which a person from birth to marriageable age is resident in her or his natal residence group, and upon marriage the spouse comes from a different residence group. Residence group marriage exogamy may sometimes be *de facto* as happens with the !Kung San who do not have explicit rules against marrying someone from the same residence group. Instead, as shown by multi-agent simulation, the kinship expressed marriage taboos lead to *de facto* residence group marriage exogamy (Read 1998). More often, it is explicit, as in hunter-gatherer groups such as the Kariera of Australia. Now consider a specific residential group G. We will assume that each person born in G is a first order kin of every other person in G and, prior to marriage, all of a person's first order kin are members of G. Although these assumptions underestimate the number of first order kin for a person by overly localizing their distribution in a single residence group, we will compensate for this underestimation by overestimating the degree of dispersal of individuals into other residential groups through marriage.

We will assume that each person in G is associated, through marriage, with a different residence group, namely the natal residence group, G^*, for a person's spouse. Each person in G^* becomes a second order kin to each person in G through the marriage link. By assuming maximum dispersal of marriage links, there will be as many other residence groups, G^*, with second order kin relations to the members of G as there are persons in G. For any other residence group, the connection to G would require a third order kin relation. Now assume, for simplicity in computation, that each residence group has n persons. Consequently, the number of first and second order kin for persons in G will be n^2. Finally, we will assume that the equilibrium pattern of the society for social interaction is that of reciprocal social interaction, hence each person in the society of interacting persons must be a first or second order kin to every other person in the society, and so the society population size of mutually recognizable kin will be bounded by n^2.

Agreement between Data and Prediction

The size of a residence group has been estimated to be 25 persons (Kelly 1995) regardless of environmental and ecological conditions, and that value has been justified by referring to the fact that consensus decision making becomes problematic when there are more than about five to seven decision makers (namely heads of families) involved (Johnson 1982), in part due to limitations on the size of working memory (Bernard and Killworth 1973). Assuming a residence group depends on consensus decision making in simple hunter-gatherer societies, and assuming that each family acts as a social unit with a single decision maker for the purposes of residence group consensus decision making, a residence group should have a maximum of five to seven families before internal conflict may lead to fissioning. For hunter-gatherer populations with an overall stabilized population size, the average family size will be around 5.2 to 5.3 persons (two surviving offspring to replace two parents, one additional offspring to account for approximately a 20 percent mortality rate during reproductive years, plus a fraction of a person to take into account biological infertility rates of 10 to 15 percent in human populations) and so the expected residence group size (rounding off to the nearest five persons) will be around 25 to 35 persons.

These estimates are consistent with data on the size of residence groups. Figure 5.7 shows the frequency distribution for the size of "the consumer group that regularly camps together during the most aggregated phase of the yearly economic cycle" (Binford 2001: 117). The heterogeneity implied by the bimodality in the histogram indicates that we should only consider residence groups with $p < 42.5$ (the midpoint for the range [41, 44] of values included under the value of 44 in the histogram). For the left mode in the histogram, $\bar{x} = 30.0$, consistent with Johnson's argument

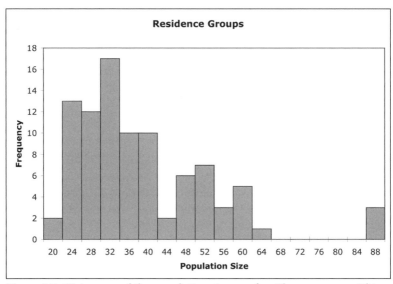

Figure 5.7 Histogram of the population size, p, of residence groups within a hunter-gatherer society. Two societies with $p > 130$ are not shown. *Data are from Binford 2001: Table 5.01.*

for residence group sizes between 25 and 35. A similar mean residence group size of $\bar{x} = 28.2$ was obtained for a group of $n = 32$ hunter-gatherer societies in widely varying ecological conditions (Hill et al. 2011). With $p = 30$ and maximum dispersal of spouses across 30 residence groups, we would expect a maximum of 900 first and second order kin. With marriages for a residence group distributed over fifteen residence groups (hence two marriages linking each pair of residence groups), we would expect around 450 first and second order kin. Consequently, the empirically derived modal value of $p = 500$ with a range up to $p = 775$ persons is consistent with the predicted range for the population size of hunter gatherer societies with reciprocally interacting first and second order kin.

A social field between 450 and 900 persons for reciprocally interacting kin is also consistent with limitations on the size of networks of interacting individuals without hierarchical structure, as is the case with hunter-gatherer groups. A model of randomly interacting individuals shows that they can collectively interact in a cohesive manner in groups of up to about 140 individuals without hierarchical structure (Bernard and Killworth 1973, 1979; see also references in Dunbar 1993). Accordingly, hunter-gatherer groups structured through families acting as social units, with an average family size of about 5.2 persons, should have a maximal social field of around (5.2 persons per family) × (140 families) = 728 persons, well within the estimated range of a social field based on kinship connections.

This implies that the social field determined through kinship relations corresponds to a maximal cohesive social field from the viewpoint of behavior, thus simple hunter-gatherer societies are likely to exhibit stable modes of adaptation under most environmental conditions. The exceptions would be regions where wild food resources enable an order of magnitude or greater increase in population density through factors such as labor intensification in resource procurement. Conditions like this would include the northwestern parts of North America with massive salmon runs, highly abundant acorns in oak groves in the southwestern United States, and the southern California and southwestern Florida coasts, each with large quantities of shell fish. All of these are regions where complex hunter-gatherer societies developed. Similar food resource properties would also characterize the domesticated food resources that were integral to the Neolithic revolution.

We obtain further corroboration that the population size of simple hunter-gatherer societies is constrained by the internal dynamics of

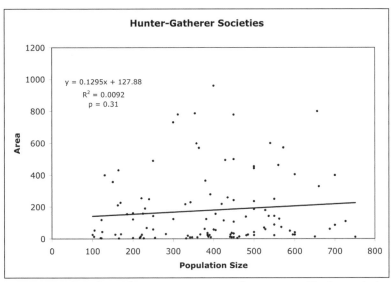

Figure 5.8 Plot of population size versus catchment area. Twelve societies with population size < 100 and three with area > 1700 square miles have been excluded as anomalous cases. *Data are from Binford 2001: Table 5.01.*

social organization rather than by external ecological and environmental conditions through the former predicting that the area for a hunter-gatherer society should be independent of population size. The prediction follows by noting that the population size distribution across different ecological zones will be the same, keeping fixed the area, when population size is constrained just by the dynamics underlying the formation of first and second order kin. This implies that population size and area should be statistically independent variables. Data on hunter-gatherer area and population size corroborate the prediction, as statistically there is no relationship between population size and area ($r^2 = 0.01$, probability $p = 0.31$; see Figure 5.8).

I conclude, then, that for hunter-gatherer societies without political structure and consisting of reciprocally interacting first and second order kin, the residence group size will have a modal value of $p \simeq 30$, and that such societies will have a modal population size of $p \simeq 500$ as originally proposed by Birdsell (1953). Large hunter-gatherer populations ($p > 775$) therefore reflect a more complex form of social organization than do residence groups based just on socially interacting first and second order kin.

Summary

The evolutionary trend in social relations and social organization when going from the OW monkeys to the great apes does not extend forward in time to hunter-gatherer societies in a direct way. Instead, the social complexity introduced by increased individualization of behavior eventually led to a shift from face-to-face to relation-based systems of social organization. This shift involved implementation of systems of social interaction in which the acting individual takes into account the potential response of the responding individual to his or her actions. Familiarity gained through intensive face-to-face interaction is required by chimpanzees to implement this kind of behavior.

The transition from experiential-based interactions to relation-based interactions depended on the exploitation of several cognitive abilities, only some of which are found in the cognitive repertoire of chimpanzees. These include a concept of self, a theory of mind, relation based categorization of dyads of individuals, the concept of reciprocal relations, and the ability to construct new relations through recursion. The limited size of working memory among chimpanzees (Read 2008c) accounts (in a proximal sense) for the absence in the chimpanzees of cognitive abilities such as recursive reasoning and conceptualization of reciprocal relations, as well as, possibly, the lack of a full-fledged theory of mind.

As long as social systems depended on face-to-face interaction for working out the social and behavioral relations among individuals, social units were composed of individuals who interacted, or had the potential for interacting, on a daily basis. Chimpanzees were able to expand the average size of communities to approximately twice the average size of troops of OW monkeys, but at the cost of reducing the internal cohesion of a community. Simultaneously, selection for more individualistic behaviors has led to a substantial increase in social complexity. Correspondingly, the neocortex ratio increases with group size when comparing OW monkey troops to chimpanzee communities.

More generally, encephalization and expansion of the size of working memory during the evolution leading to *Homo sapiens* enabled a different mode of social organization to be introduced based on categorization through commonality in the relation between pairs of individuals rather than commonality in the attributes of individuals. From an initial set of relations, a more extensive system of relations could be derived by constructing new relations from the initial relations through a theory of mind projection and recursive composition of relations. Theory of mind makes it possible for one individual to perceive not only a relation (such as the mother relation) that he or she has to another individual, but simultaneously to conceptualize that the other individual has the same relation to yet a third individual. Recursive composition enables conceptual formation of a new relation (in this case the mother of mother relation). Critical to this process of forming a new relation from already conceptualized relations is the fact that implementation of the new relation as a way to categorize pairs of individuals does not depend on prior interaction between those individuals as the basis for the categorization. A female may be perceived as satisfying the mother of mother relation, for example, simply by forming the mother of mother relation through recursive composition of the mother relation with itself.

Recursive composition of relations, starting with relations that identify positions within families, underlies the way we trace genealogical relations among individuals. The conceptual ability to form new relations through recursive composition is fundamental to small scale human societies—ancestral to all modern-day societies. These small scale societies are, and have been, organized around systems of culturally constructed kin relations. The system of kin relations is expressed concretely through the conceptually linked kin terms that make up a kinship terminology as discussed in chapter 3.

The system of relations constructed from a few initial relations referring to family positions incorporates reciprocity as an integral property. If individual A has a mother relation to individual B, for example, then B conceptually has a relation to A, namely what we call a child relation, and is of the form "B has A as an individual for whom B has the mother relation to A." Reciprocity ensures that kinship relations are mutual when reciprocity is incorporated into a system of kin relations. Reciprocity ensures that the kin relation of one person to another entails a kin relation from the second person back to the first person. In English, if she is my aunt, for example, then we both understand that I am her nephew (or niece, depending on my sex).

Kinship relations incorporate expected patterns of behavior as identified by Fortes in his kinship Axiom of Amity. For example, the mother relation attributed to the macaques derives from categorization based on altruistic female parenting behaviors directed towards biological offspring. The altruistic behaviors are positive and supportive from the viewpoint of the recipient, hence the mother relation has positive and supportive behaviors associated with it. When the behaviors associated with a relation are also carried over to the composition of relations, then the mother's mother relation would, for example, also carry with it the expectation of positive and supportive behavior. More generally, since biological kin selection can select for altruistic behaviors directed towards biological family members, a system of relations constructed through composition of family relations should have associated with it positive and supportive behaviors as asserted by Fortes in his kinship Axiom of Amity. In other words, once something like altruistic behavior or cooperation becomes part of the primary relations upon which a system of relations is constructed, then we no longer need to account for the behavior in isolation via biological kin selection, but can do so instead by reference to that system of relations. Thus, to account for altruistic behavior among those who are part of a social group, it suffices to refer to the conditions under which that social group is based on those who are cultural kin to one another and to the way altruism, initially directed towards close biological kin via biological kin selection, then becomes associated with cultural kin relations through the composition of family relations among which altruism is already expressed. The shift from selection for a behavior at the individual level to associating that kind of behavior with cultural kin implies that the behavior may occur essentially independently of biological affinity, as is the case for hunter-gatherers.

When reciprocal relations are part of the system of relations and there are expected behaviors associated with relations, the basis for social interaction is in place. Through theory of mind projection, individual A perceives that the relation A has with individual B as a target makes A the target of the reciprocal relation from B's perspective. Thus, individual A believes that B will direct behavior b towards A because of B (according to A's belief) having A as a target for the reciprocal relation. Consequently, individual A directs behavior b towards B in the belief that B will direct behavior b towards A. When individual B also has the same system of relations as part of her/his conceptual repertoire, then the conditions for symmetrical social interaction are in place, since B constructs a similar scenario regarding B's belief that A will direct behavior b towards B and so directs behavior b towards A in the belief that A will direct behavior b towards B. Critically, when A and B share the same conceptual repertoire, symmetric social interaction will be reinforced since each of A and B direct behavior b towards the other individual, thereby reinforcing the beliefs by A and B regarding the likely behavior of the other individual. Sharing the same conceptual repertoire becomes a criterion for the reinforcement of symmetric social interaction. This kind of sharing becomes the basis for cooperative behavior directed towards those individuals who simultaneously are the target of the relations in a system of relations and have that system of relations as part of their conceptual repertoire.

The social field is determined, then, not by boundaries on face-to-face interaction, but on those individuals sharing the same conceptual repertoire. This implies that the scope of the social field is determined through enculturation; that is, by the collection of individuals whose interaction leads to offspring raised understanding the same conceptual repertoire of relations. With kinship systems developed as an elaboration of the relations derived through female parenting and male-female pair-bonding-like behavior (as a way to ensure female access to high-nutrition foods obtained through scavenging and hunting when females became less mobile due to the secondary altriciality of offspring) culturally constructed kinship relations became the means to identify those enculturated into the same cultural repertoire. This also implies that the size of the social field will be determined by the number of first and second order kin, which is on the order of n^2, where n is the number of first order kin. With residence groups of around 30 persons, this implies that the size of the social field should be around 675 persons, the midpoint of the estimated range of 450 to 900 persons for the range of interacting kin, regardless of ecological conditions. The estimate of 675 persons is in accord with data on the size of hunter-gatherer groups.

Notes

1   Compare with the definition of a solitary primate species as one in which "the general activity, and particularly, the movements of different individuals about their habitat are not synchronised" (Charles-Dominique 1978: 139, as quoted in Kappeler and van Schaik 2002: 710).

2   Although both the orangutans and some of the prosimian species have forms of social organization characterized as solitary but social, the two primate groups differ substantially with regard to degree of individualized social behavior, suggesting that they may be solitary for different reasons. Solitary prosimians are nocturnal, and the absence of a group structure appears to be an effective response to low density of resources and/or low risk of predation (Keppler and van Schaik 2002). Orangutans, however, may be coping with high individuality in social behavior by minimizing the size of a social unit. Social units formed for reasons relating to mating and reproduction did occur on about two-thirds of the observation days of focal individuals among the more gregarious orangutans in Sumatra and had a mean of 2.5 individuals per non-solitary social unit (extrapolated from data in van Schaik 1999). In addition, male orangutans form dominance hierarchies that are acted upon when there are chance encounters (Cawthon Lang

2005, and references therein). Together, these observations suggest that the solitary behavior of the orangutans may be due to a reduction in the size of social units and not, for example, an adaptation to resource distribution and/or predation conditions.

3   Symmetric social interaction is not the same as reciprocal altruism. Reciprocal altruism has to do with hypothesized conditions under which there may be selection for altruistic behavior through biological kin selection. Social interaction has to do with learning and assessing the likelihood of behaviors rather than biological kin selection, and so, unlike the situation with altruistic behaviors, no fitness cost need be assumed with social interaction behaviors.

4   Individuation is being used here to characterize observable differences in behavior that arose through evolution under particular conditions and not, as Spencer expressed it, that evolution has a "tendency to individuation" (Spencer 1851: 456).

5   More formally, by the recursive composition of two relations, **R** and **S**, in that order, over a set S of individuals we will mean the relation consisting of all dyads that can be formed in the following manner. Start with an individual A in S as the focal person for the **R** relation, then find an individual B (if any) in S such that the dyad (A, B) satisfies the **R** relation. For such a dyad, take the person B as the focal person for the **S** relation and find a person C (if any) in S such that the dyad (B, C) satisfies the **S** relation. The dyad (A, C), if it exists, along with all other dyads formed in this manner, makes up the relation that is the recursive composition of **R** and **S**.

6   When **R** and **S** are reciprocal relations, the composition relation **RS** has the property that whenever the pair of individuals (A, B) satisfies the **R** relation, then the pair of individuals (A, A) satisfies the **RS** relation. This follows from noting that (A, A) will satisfy the **RS** relation when there is an individual B such that the pair of individuals (A, B) satisfies the **R** relation, then the pair of individuals (B, A) satisfies the **S** relation. If **R** and **S** are reciprocal relations, then whenever (A, B) satisfies the **R** relation, (B, A) satisfies the **S** relation given what it means for two relations to be reciprocal relations. Now let **R** be the parent relation and **S** the child relation. For every individual A there is an individual B such that (A, B) satisfies the parent relation, hence for every individual A, the pair (A, A) satisfies the composition of the parent and child relation. For this reason we can call the composition of the parent and the child relation the identity relation; that is, the relation that associates an individual A with her/himself.

7   A conceptual linkage between individuals is not universally a necessary prerequisite for social interaction, as shown by the eusocial insects, if we do not require awareness to be part of anticipated behavior.

CHAPTER 6

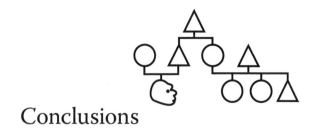

Conclusions

Odyssey Beginning Point: Old World Monkey Societies

In the odyssey I have outlined in the preceding chapters, the OW monkeys provide us with a beginning point in which social organization and structure are explicable by reference to biological evolution driven by natural selection and biological kin selection. Within this context, social organization for the OW monkeys appears to be a balance between the opposing effects of predation risk leading to larger groups and resource competition at the level of individuals having the opposite effect, taking into account the way male-female relations play out in a particular species. For the female philopatric species, social organization revolves around female matrilines with biological kin selection for altruistic behaviors such as grooming. The combination of social units based on matrilines and a stable dominance relation among matriline units implies that social complexity scales with the number of matriline units, not the total number of females. Correspondingly, the neocortex ratio, taken as a measure of social complexity, differs systematically between modes of adaptation such as arboreal versus terrestrial foraging, and only indirectly with troop size.

Odyssey Endpoint: Hunter-Gatherer Societies

Our ending point of cultural kin-based, hunter-gatherer societies without political structure or hierarchical institutions contrasts sharply with non-human primate forms of social organization. Here we find cultural idea systems such as rules for access and ownership of resources in lieu of the individual competition we see in OW monkey troops. Hunter-gatherer societies, unlike non-human primate species, have rules for sharing resources that equalize access to high variance, highly valued food resources such as large game animals. These cultural idea systems for sharing are not explicated by referral to a transmission mode such as direct, phenotypic trait transmittal or biological kin selection. Instead, they show the way an idea system constituted as a "complex whole" provides the framework within which individuals act and interact.

Culturally constituted kinship idea systems, expressed concretely through kinship terminologies that define who are one's kin, are central to understanding the social organization of hunter-gatherer societies. The kinship terminology provides a symbolic, computational system through which a conceptual boundary for socially interacting individuals is implemented. We visually make evident the logic of this computational system through a kin term map showing the conceptual relationships among the kin terms. The kin term maps for two hunter-gather societies, the !Kung san of Botswana and the Kariera of western Australia, show marked differences in their respective structural organization of kin despite similar modes of ecological adaptation. In other words, the kinship system is not simply the epiphenomena of behaviors adapted to environmental and ecological constraints.

Cultural idea systems provide the conceptual basis and rules for ownership and sharing of resources. Resources in the wild are collectively owned by a group, and access to them is either through group membership or by culturally defined kinship ties with members of the group. Except for a family context where sharing is universally expected, cultural rules for individual ownership negate any obligations for sharing resources. Individual ownership is associated with resources that are equally obtainable by all able-bodied adults. Collective ownership, however, along with rules for sharing, occurs with resources that come in large units (such as game animals) not equally obtainable by any able-bodied adult.

Hunter-gatherer societies are egalitarian in that they lack hierarchical institutions and maintain an ideology of males and females acting for the benefit of the group. Failure or refusal to follow rules for sharing resources implies that one is opting out of adherence to a cultural idea system through which one's identity as a social person is defined. Opting out of conformity with a cultural idea system is not just an individual act, since it challenges all of those who share the same cultural idea system. The response of those who have been challenged may be to use social pressure to get those they perceive as deviant individuals to conform to the cultural idea system.

Odyssey Midpoint: Chimpanzees

Chimpanzees provide a midpoint for this odyssey, and so we can compare the trend going from the OW monkeys to the chimpanzees to the one going from chimpanzees to our species, *Homo sapiens*. Regarding the first trend, the chimpanzees differ from OW monkeys by the increased emphasis on their performative (hence emergent) rather than ostensive (hence structured) forms of social organization. The organization of chimpanzee communities emerges out of intensive interaction between individuals, with individualistic behavior on a scale comparable to that of humans. The transition from OW monkeys to the great apes thus incorporates increased individualization of behavior, which leads to greater complexity of social relations among individuals. This trend can also be seen in the distinctive form of social organization for the major taxa of the greater and lesser apes: pair-bonding for the hylobatids, solitary behavior for the orangutans, harem social units for the gorillas, and multi-male and multi-female communities for the chimpanzees. That there are different forms of social organization among taxa is mirrored at a smaller scale by variation in social behaviors among chimpanzees from different communities. In addition, social units within a community composed of adult males are small and unstable. Instability also characterizes dominance hierarchies of males, as these are performatively constructed and subject to challenge. Biological kin selection appears to be less important than is the case for the OW monkeys, as male behaviors are not biased towards paternal genetic kin despite male philopatry.

Social Complexity Through Individualistic Behaviors

Social groups are more complex in the chimpanzees due to the prevalence of individualistic behavior. The general trend from the OW monkeys to the chimpanzees is towards increased individualization of behavior that would lead to highly complex social systems were it not the case that social organization is restructured in the form of small social units for males and individually oriented social units for females. The complexity of the chimpanzee social organization, even taking the restructuring into account, is, as shown by their neocortex ratio, even greater than it is for the OW monkeys. Chimpanzee social organization appears, then, to be a way to accommodate increased individuality within a biologically constituted social system based on face-to-face interaction. Along with increased individualization of behaviors, we find greater prevalence of socially transmitted behaviors. When we compare chimpanzee systems of social organization with those of hunter-gatherer groups, though, we do not just find increased elaboration on the properties and features of chimpanzee forms of social organization. Instead, the key difference is a shift to relation-based systems of social organization built around culture as a "complex whole" transmitted through enculturation. Culture, in this sense, is not a sum of individual traits, each transmitted through imitation/learning. Instead, it is what is being transmitted, not the mode of transmission, that expresses the difference between chimpanzee and hunter-gatherer forms of social organization.

Kinship Terminologies as a Complex Whole

This difference is particularly clear in the kinship terminologies of hunter-gatherer groups as these express a logically consistent system of concepts and not simply a list of kin terms that label categories of genealogical relations. Logical consistency assures coordination across individuals in what is shared, and this logical consistency is made visually evident in a kin term map. Logical consistency underlies the kin terms as a system of symbols through which kinship relations may be symbolically computed. In this way, the terminology defines who are kin to whom by way of kin relations computed in a manner consistent from the perspective of each individual who has the terminology as part of his or her cultural repertoire.

The terminology provides structure and organization for the domain of culturally defined kin. This implies that we need to distinguish between behaviors made according to biological/cognitive information, on the one hand, and cultural information on the other hand (see Figure 4.6). Unlike biological/cognitive information, which can be obtained by an individual in isolation, cultural information is embedded in the individuals making up a social unit and from whom the information is obtained through enculturation. A child becomes knowledgeable about a kinship system and its terminology through interaction with other, already knowledgeable individuals, not through individual learning in isolation. Consequently, we need to distinguish between sharing of cultural knowledge stemming from transmittal of a cultural idea system through enculturation, on the one hand, and sharing that just represents the statistical frequency for the distribution of an individually learned or acquired trait among the members of a group of individuals on the other hand.

Cognitive Basis for the Expansion of the Social Field

The transition from experiential-based to relation-based social behavior depends on access to cognitive abilities, such as a concept of self, a theory of mind, relation based categorization of dyads of individuals, the concept of reciprocal relations, and recursive composition of relations. Implementation of these cognitive abilities depends on having a working memory greater than is found in chimpanzees. The more limited working memory of chimpanzees ensures that these cognitive abilities will only be exhibited among the chimpanzees in restricted form at best. Hence, the odyssey leading to *Homo sapiens* was not, and is not, open to the chimpanzees, given their mode of adaptation.

Of these cognitive abilities, theory of mind provides the basis for the composition of relations when one individual conceptualizes that another individual has the same relation to a third individual as the first individual has to the second individual. Recursive composition makes it possible to form a new relation from already conceptualized relations without first requiring prior interaction between the pairs of individuals categorized according to the composition of relations. A mother of mother relation, for example, can be formed through recursive composition of the mother relation with itself without requiring that there is already a regular pattern of behavior between biological grandmother and biological granddaughter pairs upon which the mother of mother relation is based.

Tracing genealogical relations between pairs of individuals uses recursive composition of relations and incorporates reciprocal relations. If individual *A* has a mother relation to individual *B*, then *B* conceptually has a child relation to *A* of the form "*B* has *A* as an individual for whom *B* has the mother relation to *A*." Reciprocity ensures that kinship relations are mutual since reciprocity is incorporated into a system of kin relations. Reciprocity ensures that the kin relation of one person to another determines simultaneously a kin relation from the second person back to the first person.

Kinship relations incorporate positive, expected patterns of behavior. The mother relation attributed to the macaques, for example, derives from categorizations based on altruistic female parenting directed towards biological offspring. The altruistic behaviors are positive and supportive from the viewpoint of the recipient, hence the mother relation has positive and supportive behaviors associated with it. When the behaviors associated with a relation are also carried over to the composition of relations, then the mother's mother relation would, for example, also carry with it the expectation of positive and supportive behavior. More generally, since biological kin selection will favor altruistic behaviors directed towards biological family members, a system of relations constructed through composition of family relations should have associated with it positive and supportive behaviors.

When reciprocal relations and associated positive, expected behaviors are part of the system of relations, the basis for reinforced social interaction is in place. Theory of mind projections and reciprocal relations imply that the relation A has with individual B enables A to formulate the belief that A is the target of the reciprocal relation from B's perspective, and so A believes that B will direct the kind of behavior associated with a relation towards A. When individual B has the same system of relations as part of her/his conceptual repertoire as does individual A, then symmetrical social interaction will be reinforced. When each of A and B act according to her or his respective beliefs about the other individual, each of A and B is the recipient of behavior by the other individual—behavior that reinforces each individual's beliefs. This implies that sharing the same conceptual repertoire leads to the reinforcement of symmetric social interaction, and hence leads to reciprocated cooperative behavior directed towards those individuals who have the same system of relations as part of their conceptual repertoire.

The outcome is a social field determined by those individuals sharing the same conceptual repertoire, hence the social field is determined by those individuals enculturated with the same system of relations. Being kin to one another through culturally constructed kin relations becomes a means to identify those enculturated with the same cultural repertoire. This leads to the use of kin relations to determine those with whom social interaction may take place. The size of the social field determined in this manner will be determined by the number of first and second order kin. As a result, the size of the social field should be around 625 persons based on residence groups of 25 persons, regardless of ecological conditions—a magnitude in accord with data on the size of simple hunter-gatherer groups.

Implications of Evolutionary Change in Social Systems

The odyssey I have traced from our primate ancestry to the beginnings of the hunter-gatherer societies that form the crucible for the development of the more recent, larger scale societies has behavioral ramifications that have only been partially fleshed out through research. This contrasts with the morphological side of the odyssey wherein the anatomical changes that separate us from our primate ancestors have been worked out, at least in broad strokes. From a morphological perspective, though the number of hominin individuals represented by the fossil material that has been recovered is still small and for some time periods is rare or nonexistent, we have a fairly good idea of the time-based pattern of anatomical changes that took place in the evolution of our species, *Homo sapiens*. Nonetheless, unexpected and unanticipated regional variants are still found, such as the extinct *Homo florensis*, the small hominin found in 2004 on the Flores Island in Indonesia, or the Neanderthal-like hominin material found in 2010 in a cave in Denisova, Siberia, with DNA overlapping that of individuals from New Guinea. Finds like these help fill out our understanding of our morphological ancestry.

Theoretically, we interpret variation in morphological characteristics of fossil material through the Darwinian account of biological evolution driven by natural selection acting on change in biological information introduced through genetic mutation. In the Darwinian account, directionality is derived from genetic mutation by natural selection acting on characteristics transmitted across generations through biological reproduction and development. The biological processes of reproduction and development ensure absence of direct feedback from the phenotype to the genetic information contained in the genome of an individual. Consequently, the apparent directionality of evolution towards increasing the morphological complexity among extant organisms through time derives from the process of natural selection acting to increase the fitness of individuals.

In comparison to morphological evolution, the dynamics of the ways that cultural and behavioral evolution relate to the structural and social organization of societies is less well worked out. But some of the differences between morphological evolution and cultural evolution are apparent. Rather than simply waiting for directionality to be determined through biological fitness selection, humans can intentionally and deliberately modify the information content underlying the structural and social organization of a society in a directed manner. Unlike the isolation of genetic information from the processes involved in the development of the physical phenotypic of an individual from a fertilized egg, the capacity for intentional feedback modifying cultural information is an integral part of a cultural system.

Despite the success of models for the evolution of morphological characteristics, the same models are insufficient when applied by analogy to the evolution of cultural systems and their attendant behaviors. This insufficiency has been partly addressed through new models of evolutionary processes that take into account the phenotypic transmittal of behaviors from one individual to another through social learning. Social learning became increasingly important, along with greater individualization of behaviors, in the phylogenetic transition from OW monkeys to the great apes (see Figure 4.5). Among the chimpanzees, we already find numerous examples of socially transmitted behaviors with community specific distributions.[1]

The importance of the non-biological transmission of behaviors in the evolution leading to our species is evident, and there has been extensive work on modeling the consequences of direct phenotype trait transmission. Although it is important to recognize that a different transmission mode underlies the behavioral differences we observe when comparing human societies with non-human primate societies, the transmission mode, by itself, does not account for the substance of the changes that took place in social systems leading to present-day human societies. These changes are grounded in the transition from face-to-face interaction, as the behavioral basis upon which emergent forms of social organization arise, to relation-based interaction, where it is the organization of relations that is critical and not the traits of individuals, per se.

Cultural Idea Systems as Group-Level Properties

The kinship boundary we find circumscribing the social field in hunter-gatherer groups does not emerge from the Darwinian process of traits introduced through mutations followed by change in trait frequency through selection operating at the level of individuals. Instead, the kinship boundary is determined by the scope for the cohort of individuals among whom the computations of a kinship terminology make it possible to determine if individuals are kin to one another (see chapter 5). The kin terms making up the terminology have functionality through the organization and structure that the terminology provides for the members of a social group. As a result, functionality of a kinship terminology depends on it being part of the conceptual repertoire of the interacting group members and not simply as traits associated with individuals. It is not the fact of an individual knowing who are his or her kin that causes the functionality of the kinship system. Rather, the functionality derives from the structural and conceptual organization the terminology provides for the members of a group, and hence for the organization of the group as a whole. The expanded social field we find in hunter-gatherer societies is made possible by the kinship relation system expressed linguistically through the kin terms making up a kinship terminology.

Kinship relation systems enable the redistribution of families in residence groups according to kinship ties and according to local variation in resource abundance, thereby substantially increasing the population density that may be maintained when the scale for spatial variation in resource abundance is comparable to the geographical scale for the distribution of residence groups. The difference in population density between groups with different systems of organization is critical. A group with a system of organization that can maintain a substantially larger population density will win out in competition with another group that can only maintain a much lower population density.[2]

A temporary transfer by a family from one residence group to another depends on kin ties between the families in question. This transfer process differs structurally from the conditions for biological kin selection where already existing interaction among biologically related individuals provides the context for the selection of altruistic, fitness-enhancing behavior directed towards close biological kin. Under biological kin selection, a trait is expressed initially at the individual level and then spreads throughout a population due to higher inclusive fitness. The temporal sequence is selection of a trait to interaction among group members.

We can see how this works out with interaction among females in an OW monkey troop. Female philopatry and the dominance hierarchy in OW monkey troops lead to troops with social units composed of matrilineally related females, hence face-to-face interaction involves biologically related females by that fact alone. Under these conditions, biological kin selection can favor individual traits such as a grooming propensity, even if it does not provide direct fitness benefit to the acting individual but instead provides an adaptive benefit to the social unit through the recipient of the grooming. The adaptive group benefit relates to the functionality of grooming in maintaining and strengthening the social cohesion of a social unit composed of biologically related females. Since the recipient of the grooming is the biological kin of the acting individual, the inclusive fitness of the acting individual is increased, and selection favors the spread of the grooming trait. Thus social cohesion in a system like this emerges through interaction among individuals stemming from the behavioral traits of individuals.

With increased individuation in behaviors, the mode for the transmission of traits expands to include transmission through imitation and copying (see Figure 4.5), but the dynamics of evolutionary change remain essentially the same: selection goes from the functionality of behaviors at the individual level to group-level functionality that emerges from individual behaviors. Formation of coalitions among male adults need no longer be dependent on, nor an expression of, biological kin selection, but the selective process is still essentially the same. Traits that are first initiated at the level of individuals may increase in frequency through imitation or copying, thereby possibly leading to group-level functionality as an emergent phenomenon. Nut cracking by chimpanzees would be an example, since it does not appear to be a behavior under biological control (though the capacities brought into play in the action of nut cracking may be genetically grounded), hence it is likely that it has spread in the chimpanzee communities where it is found through a combination of individual learning and imitation or copying. At the group-level, a community engaging in nut cracking has the adaptive advantage of an additional resource added to its dietary mix. Like other components of the diet that are considered to be a property at the species level, or in some cases a property at the level of a unit such as an OW monkey troop or a chimpanzee community, cracking and eating of nutmeats is a community trait in that it is either engaged in by all community members (e.g., the Taï chimpanzee community [Boesch and Boesch 1983]) or at least attempted by all community members (e.g., the Bossou chimpanzee community [Matsuzawa 1994]) or by no individuals (e.g., chimpanzee communities in the eastern part of Côte d'Ivoire [Boesch et al. 1993]).[3] This contrasts with the hunting and eating of monkey prey, since even in communities where hunting occurs extensively, such as Ngogo, Kibale National Park in Uganda, females rarely hunt and not all males pursue prey (Watts and Mitani 2002). A similar differentiation, but in the opposite direction, between males and females also occurs among the savannah chimpanzees in Fongoli, Senegal, where the hunting of bush babies by using a pointed stick to immobilize them has been observed (Pruetz and Bertolani 2007). Thus, hunting by chimpanzees is an activity whose occurrence is variable at the individual level, and as such is an individual level trait from the viewpoint of selection. However, despite nut cracking having become a community-level trait, accounting for its spread throughout a community does not require reference to group-level adaptive advantage, or group selection, but only to individual-level adaptive advantage due to a nutritious food being obtained through the behavior.

I distinguish between the group-level adaptive advantage that may arise from emergent group-level properties and group-level properties that are part of what I refer to as cultural group selection. Cultural group selection, as the term is used here, does not refer to selection acting on individual or emergent group-level properties, but to selection acting on cultural properties that are inherently at the group level.[4] With group-level properties emergent from those expressed at the individual level, the unit of selection is still the individual, since a trait does not change its status as an individual-level trait just because of the emergence of group-level properties from individual behaviors. Nonetheless, the environment for the trait is changed by emergent properties in a way not expressed simply by a change in the frequency of the behaviors underlying the emergent property, and so there will be group-level feedback on individual traits. For example, as discussed in chapter 2, the stable linear dominance hierarchy of females in an OW monkey troop appears to be an emergent property arising, in part, from behaviors such as the support a female provides her offspring in agonistic encounters.

Simply knowing the frequency of females providing support for their offspring would miss the implications the linear dominance hierarchy has for selection operating at the individual level with regard to the range of response behaviors of an individual female to females outside of her matrilineal unit. The ongoing system with a linear dominance hierarchy is part of the environment to which individuals adapt. However, the support behavior by females for their offspring as a causal factor leading to the dominance hierarchy is not a consequence of group selection, since the dominance hierarchy would not exist without those behaviors already being in place. Thus, while the support behavior provided by a female to her offspring has group-level consequences, its frequency of occurrence is not due directly to group selection. With cultural idea systems, however, we have group-level properties and, hence, the conditions under which cultural group selection may operate, leading to evolutionary change in cultural idea systems.

Shift from Individual to Organizational Evolution

The distinction between cultural idea systems with group-level properties and individual-level properties leading to group-level, emergent properties can be seen in the system of sealing partners among the Netsilik discussed previously. The system of sealing partners had group-level functionality from which individuals received fitness benefits rather than being a system emerging from, say, the fitness benefits of individual behavior, such as food sharing among family members, that occurs even in the absence of cultural rules. The fitness benefits gained by being a sealing partner depended on functionality achieved through collective conformity to appropriate behavior, and were reinforced by a series of dyadic relationships such as song partners that allowed for "latent ambivalences to find a harmless outlet" (Balikci 1970: 141). Overall, sealing partners were a cultural group-level trait giving rise to fitness benefits for individuals embedded in the system of sealing partners, and not the reverse.

As a group-level trait, the cultural system of sealing partners was interlocked with other culturally determined group-level traits that jointly constituted an effective, adaptive system for the extreme arctic conditions with which the Netsilik and other Inuit groups had to cope. Specific hunting behaviors—such as hunting seals thorough breathing holes in the winter pack ice, forcing caribou into lakes and then killing them from kayaks during the caribou's fall migration south, obtaining large quantities of salmon during the salmon runs in the summer to be stored as food when other food resources were not available, and an adult sex ratio skewed towards male hunters by not allowing some newborn females to survive—are the material expression of a food procurement system well-adapted to extreme Arctic conditions.

These behaviors were culturally implemented into a cohesive organizational system—robust in some aspects and resilient in other aspects—discussed in detail in Read (2006). Briefly, in addition to the system of sealing partners, culturally determined group-level traits included the belief that a newborn only becomes human when it is named (thus abortion-like decisions could be implemented after birth and before naming a newborn without violating their moral sanctions against murder, as a way to reduce the risk of starvation by increasing the hunter/consumer ratio) and a belief in preferred (cultural) cousin marriage. Preferred cousin marriages ensured that the extended (male-based) family for each of the bride and groom was already linked by sibling relations at the parental level. When the linking parental relationship was that of either two brothers or a brother and a sister, the extended families of the bride and groom already had close kin connections. When the linking relationship was two sisters and their respective husbands did not have close kin ties, the extended families were then linked through another cultural institution, the *angayungoq-nukangor* ('brother-in-law') relationship between the husbands of the two sisters (Balikci 1970). In all cases, then, cousin marriage acted to reinforce already existing cultural kin ties between extended families, and the extended families functioned as groups of cooperative and sharing kin.

An extended family, possibly including another closely linked extended family, provided the basic labor pool for salmon fishing in the summer and caribou hunting in the fall. Sharing took the form of "generalized reciprocity" (Sahlins 1965: 167) in which no calculation was made of quantities or whether sharing was balanced (Damas 1972). The Netsilik and other central arctic Inuit distinguished this kind of voluntary meat sharing, which they referred to as *payuktuq*, from *niqa-iturvigiit* (sharing based on sealing partners) (Damas 1972). *Payuktuq* sharing was considered to be proper behavior among the close kin who made up the members of an extended family. Sharing food like this through generalized reciprocity among one's close kin continues to the present day in the descendant Inuit groups in this region (Collings, Wenzel, and Condon 1998).

Seal hunting could not rely on voluntary sharing among closely related families, as the number of males required for averaging out the risks when hunting seals through their breathing holes in the pack ice required a larger cohort of hunters than would be found in the extended families connected through marriage and other kin ties. Sharing among sealing partners is an example of Sahlins's notion of balanced reciprocity (Damas 1972) and had the functionality of reducing conflict by defining precisely the obligations and rights of sealing partners with respect to the meat and blubber from a seal killed by one of the winter camp hunters.

Hunting seals through the pack ice probably originated in the sixteenth century AD in response to the maximum cold period of the 'Little Ice Age.' Previously, whale hunting had been central to the Thule Inuit, the direct ancestors of the present-day Inuit, as part of their system of food procurement. The Thule Inuit migrated, starting in the ninth century AD, from Alaska into the central and northern Arctic areas during a much warmer period. With the onset of the Little Ice Age, whaling was no longer possible in the central and northern Arctic due to the extensive spread of sea ice and likely shifts in the migration patterns of whales, eventually leading to hunting of seals through the winter pack ice.

While seal hunting through breathing holes may have begun initially as an individual or small group adaptation as the winter pack ice became more prevalent, at some point there would have been a phase shift in which seal hunting as an individual-level adaptation was no longer feasible due to the increasing risk of hunting failure with the further development of the pack ice and an increase in the length of time it was present. The phase shift would have been from the initial hunting by small groups responding to changing ecological and environmental conditions to hunting by large groups dependent upon the organization provided by the system of sealing partners and other cultural systems that became the Inuit means of coping with the harsher conditions that developed in the central and northern Arctic regions.

The environmental conditions elevated the degree of risk involved in surviving under harsh Arctic conditions and thereby constrained the range of possible adaptive behaviors, thus acting as a selection device for what were viable group-level and culturally constituted means for implementing those behaviors. Overall, the adaptation of the Inuit was to culturally implemented systems of organization incorporating a variety of enabling cultural traits, such as the beliefs associated with the naming of a newborn, preferred cousin marriage (at least among the Netsilik), the extended family as a social unit (though not among the Copper Inuit [Damas 1972]), the 'brother-in-law' relationship among the Netsilik, the *payuktuq* system of generalized sharing, and, of course, the *niqaiturvigiit* system of sealing partners (Read 2006). This ensemble of cultural systems provided the organizational framework that made for group-level, rather than individual-level, adaptation. Individual fitness benefits then arose from being part of this organizational system that made it possible to occupy the Arctic under harsh and risky conditions.

Risk and Group-Level Selection

The degree of risk in obtaining food resources varies across hunter-gather groups in accord with different environmental conditions. Risk relates to the proportion of days a resource is available, the chance of finding a resource on a day when it is available, and the likelihood of obtaining it once it is found. Generally speaking, vegetal resources are available throughout the growing season, are likely to be located, almost certainly can be obtained when located, and hence are low risk resources. Animal resources are more variable regarding when they are available, are less likely to be found even when available, and attempts to kill an animal once it is found may not succeed, hence, they are much riskier resources to obtain.

A primary factor in the availability of resources is the length of the growing season, making it a useful proxy measure for risk. Hunter-gatherers can affect the availability of resources through seasonally

Table 2. Residence of Hunter

Environment: Son Resident with:	Warm	Cold
Father	3 (37.5%)	17 (63.0%)*
Mixed Pattern	3 (37.5%)	4 (14.8%)
Wife's Parents	2 (25%)	6 (22.2%)*
Total	8	27

Note: Characterization of environment and data are from
Marlowe 2004: Tables 1, 2
* Difference in per cent of son's residence with father and
  with wife's parents in cold environments is statistically
  significant at the $\alpha = 0.05$ level ($p < 0.02$)

Table 3: Residence Group Dominated by Male Relatives Versus Risk

Male Relatives:	Risk:	Low (Growing Season 12 months)	High (Growing Season 0–2 months)
Statistically significant proportion of Male Relatives ($p < 0.05$)		3 (37.5%)	17 (63.0%)*
Statistically non-significant proportion of Male Relatives		3 (37.5%)	4 (14.8%)
Total		17	8

Notes: (1) Data are from Hill et al. 2011: Table 1
(2) Read 2008b discusses the growing season as a proxy for risk
* Difference in per cent Statistically Significant Proportion in High Risk
  environments is statistically significant at the $\alpha = 0.05$ level ($p < 0.03$)

adjusting the location of residence locations and/or hunting camps. Success in locating resources relates to developing skills and knowledge, both individual and collective, regarding the location and seasonality of resources. The likelihood of success in obtaining mobile resources can be positively affected by the design for implements used in the procurement of a resource (Torrence 2001). The positive relationship between risk and design complexity (Torrence 1989; Read 2008b) implies that risk can be measured *post hoc* by the complexity of the implements used for food resource procurement. Thus the Inuit, dependent upon hunting and fishing and living in a high risk environment due to highly constrained options for hunting and fishing, made complex implements that increased the likelihood of success in each episode of hunting or fishing. With decreasing risk, the complexity of implements decreases, and there are more options for alternative forms of resource procurement. In tropical environments implements are simple, implying low risk, and groups have multiple options for obtaining food resources.

At the group level, in high risk, cold environments hunter-gatherers are dependent on cooperative hunting, typically associated with closely related males, and hence should have forms of social organization favoring sons residing, in the long term, with fathers, as argued by Julian Steward (1936, 1955) and Elman Service (1962). Under equilibrium conditions, hence with a stabilized population size, differences between forms of social organization will be expressed through group properties, such as the population density that may be maintained, and not through individual properties, such as relative fitness, since average family sizes will be the same when the population size is stabilized. Under these conditions, there will be cultural group-level selection (see Read 1987) acting on one group's culturally constructed system of social organization in comparison with that of another group with a different form of social organization. Under cold conditions with a short growing season and high risk, group-level selection will favor forms of social organization with sons residing with fathers. In low risk, warm environments, material constraints are weaker and the residence patterns should be more varied.

This prediction is supported by two data sets. First, data on hunter-gatherer long-term residence patterns show that in warm environments with low risk there is no preferred residence pattern (see Table 2), but in cold climates with high risk, sons most often reside with father (63 percent of the cold environment cases). Second, data on the statistical significance of the proportion of close male relatives in a residence group are similar (see Table 3). In low risk environments there is no difference in whether males are residing with close male relatives or not. In high risk environments, most often (88 percent of the cases) males are residing with their close male relatives.

Summary

The odyssey that I have described has had the long term consequence of shifting evolution from being executed through traits of individuals to a new mode of evolution in which the evolution of organizational systems is central: "Almost all of what we human beings have achieved ... [has] been accomplished by organizations.... The capability of human sociocultural organizations to innovate [i.e., to evolve] depends on the representations, rules, relationships, management processes and functions associated with these organizations, which are different from, and have vastly more transformative and generative capability than, those at the individual level" (Lane et al. 2009). The transition in the basis for social organization that is central to this odyssey has led to a new social order that transcends the limitations of social systems based on face-to-face interaction. New modalities of organization have been made possible in which the social order is no longer the result of individual behavior evolving to an emergent form of social organization, but a social order in which "the social group ... was freed from the constraints imposed by the conditions that individual learning through face-to-face interaction impose on social interaction.... A social group can take on functionality far exceeding the forms of social organization available to the non-human primates. New functionality could now be introduced through change in the organizational basis for societies, as expressed through change in cultural resources and tested through cultural group competition" (Read, Lane and van der Leeuw 2009: 82). In other words, we better understand human social systems by reference to the organizational systems built around cultural ideas and the implication these systems have for the adaptation of groups to their social and ecological environments than by reference to individual traits and to emergent properties arising through the interaction of individuals.

The kinship idea systems that are central to hunter-gather societies are coupled with what Fortes referred to as "prescriptive altruism," namely that kin—where kin must be understood as a cultural construct—are expected to act altruistically to each other, not because of the evolution of altruism as a biological trait, but through culturally expressed concepts of kinship that define for individuals what it means for them to be kin to one another. These culturally constructed kin relations enabled the social field to be expanded beyond the residence group, thereby leading to the transition from face-to-face to relation-based systems of social organization. This transition accounts for the crucial shift in human evolution to a social field that, even at the level of the residence group, is neither bound by biological criteria nor is it a social system dependent for its organization upon daily interaction among all members of the social field. This decoupling from biological criteria can be seen empirically in the fact that residence groups in hunter-gatherer societies are made up of individuals with only a low degree of biological relatedness (Hill et al. 2011). This, in combination with the prescribed altruistic behavior that is part of kinship cultural idea systems, accounts for both the prevalence of cooperative behavior among non-biological kin in hunter-gatherer societies discussed by Kim Hill and coworkers (Hill et al. 2011) and why it occurs primarily under culturally specified conditions. We no longer need to account for behaviors like this by models based on selection at the individual level. Instead, once the behaviors are associated with something like a cultural kinship system, selection now takes the form of cultural group selection for that cultural kinship system, and the behaviors are part of the functionality of the cultural kinship system being selected, for or against, at the cultural, rather than the individual, level, since the cultural kinship system is a group-level, not individual-level, trait.

Once the social field was implemented through a conceptual definition such as those who are cultural kin to each other, further modification and expansion of the social field could be, and has been, driven by change in, or introduction of, new conceptual definitions, such as a lineage in a tribal society defined as those who can trace back (depending on the particular tribal society) through either father links to a reference ancestor or through mother links to a reference ancestress. Cultural evolution, properly speaking, refers to changes in cultural idea systems, not to changes in the frequency of traits and the modes of transmission for behavioral traits. The key innovation (Read, Lane, and van der Leeuw 2009) marking the transition from non-human primate to human social systems and making us humans and not just another primate species is the introduction of cultural ideas and relation systems upon which human systems of social organization and structure are founded.

Notes

1 Not every learned behavior with distribution across the members of a group need be distributed by imitation or copying, since similarity in learned behaviors can be due to functional constraints that limit the range of what constitutes an effective, learned behavior.

2 With genetic traits, any fitness advantage over time of one allele over a competing allele leads to replacement of the second allele by the first allele. With two systems of social organization currently in equilibrium with each other, a small increase in the competitive advantage of one system over the other (measured through differences in population density that each can maintain under competition for resources) will only lead to a shift in the equilibrium population densities between the two systems, not replacement. One system will displace the other only with a qualitative increase in population density (Read 1987). For this reason, changes in forms of social organization in human societies, such as the change from band-level to tribal-level forms of social organization, are associated with at least order of magnitude changes in population density.

3 With nut cracking adding to the diet, we would expect nut cracking communities to win out against non-nut cracking communities (where "win out" may include diffusion of the cracking trait from a nut cracking community to a non-nut cracking community) when the nuts are present but are not cracked and eaten. Nut cracking, when it occurs, is a community level trait by virtue of it being engaged in either by all community members or by no community members. In this regard, the patterning of nut cracking in Côte d'Ivoire is particularly informative as nut cracking is divided between communities with nut cracking west of the N'Zo-Sassandra river and communities without nut cracking east of the river, even though nuts and suitable means for cracking them are available east of the river (Boesch et al. 1993). This pattern fits in with selection at the group level among adjacent communities on the same side of the river and differentiation between the two sides of the river that acts as a barrier for the chimpanzee communities. There do not appear to be any examples of geographically adjacent chimpanzee communities with one community engaging in nut cracking and the other not.

4 In the biological literature, group selection generally refers to an increase in allele frequency for biological traits (such as altruism) that are beneficial for the group as a whole but not for the individual, as a consequence of positive selection among groups overriding negative selection within groups. Earlier arguments assuming that there would be an increase in allele frequency whenever a trait had beneficial group outcomes (e.g., Wynne-Edwards 1962) have been shown to be invalid (e.g., Williams 1966), but the counter arguments do not discount group selection under all circumstances. Interest in group selection has been revived under the idea of multilevel selection (Sober and Wilson 1998). Much of the current controversy regarding the importance of multilevel selection centers on whether it only occurs under special circumstances, and hence has limited theoretical utility. Conceptually, multilevel selection is not restricted to altruistic-like traits and refers

to any situation where units at one level (subindividual, individual, group, metagroup, etc.) are subject to (natural) selection but the unit's constituent parts at a lower level are not (Wilson and Sober 1994). For biological traits at levels above the individual, these conditions are met with, for example, the social insects but not the asocial primates. Nor is multilevel selection restricted to biological traits. We will use the term *cultural group selection* when we are considering cultural phenomena and the unit of selection is a group. Cultural phenomena, by their nature, are expressed at the group-level. By this is meant that cultural phenomena cannot be fully measured or identified by reference to the individual level alone. A cultural kinship terminology, for instance, is a group-level phenomenon in that the terminology system is not reducible to just the kin term knowledge one individual may have since the functionality of the terminology system depends on the terminology being shared as part of the enculturation of the group members and is expressed through the kinship relations and expected behaviors occurring among those individuals for whom the kinship terminology is part of their enculturation.

References

Aiello, L. C., and R. I. M. Dunbar. 1993. Neocortex size, group size and the evolution of language. *Current Anthropology* 34:184–193.

Aiello, L. C., and P. E. Wheeler. 1995. The expensive-tissue hypothesis: The brain and digestive system in human and primate evolution. *Current Anthropology* 36:199–221.

Alia, V. 2007. *Names and Nunavut: Culture and identity in the Inuit homeland.* Oxford: Berghahn Books.

Aureli, F., and C. M. Schaffner. 2007. Aggression and conflict management at fusion in spider monkeys. *Biology Letters* 3:147–149.

Aureli, F., C. M. Schaffner, C. Boesch, S. K. Bearder, J. Call, C. A. Chapman, R. Connor, A. Di Fiore, R. I. M. Dunbar, S. P. Henzi, K. Holekamp, A. H. Korstjens, R. Layton, P. Lee, J. Lehmann, J. H. Manson, G. Ramos-Fernandez, K. B. Strier, and C. P. van Schaik. 2008. Fission-Fusion dynamics: New research frameworks. *Current Anthropology* 49:627–654.

Baddeley, A. D. 2001. Is working memory still working? *American Psychologist* 56:851–864.

Balikci, A. 1970. *The Netsilik Eskimo.* New York: Doubleday.

Ballerini, M., N. Cabibbo, R. Candelier, A. Cavagna, E. Cisbani, I. Giardina, V. Lecomte, A. Orlandi, G. Parisi, A. Procaccini, and M. Z. Viale. 2008. Interaction ruling animal collective behavior depends on topological rather than metric distance: Evidence from a field study. *Proceedings of the National Academy of Sciences (USA)* 105:1232–1237.

Bard, K. A., B. K. Todd, and C. Bernier. 2006. Self-awareness in human and chimpanzee infants: What is measured and what is meant by the mark and mirror test? *Infancy* 9:191–219.

Bar-Yosef, O., and S. L. Kuhn. 1999. The big deal about blades: Laminar technologies and human evolution. *American Anthropologist* 101:322–338.

Barrett, L., and S. P. Henzi. 2002. Constraints on relationship formation among female primates. *Behaviour* 139:263–289.

Barrett, L., T. Weingrill, J. E. Lycett, and R. A. Hill. 1999. Market forces predict grooming reciprocity in female baboons. *Proceedings of the Royal Society of London B* 266:665–670.

Barton, R. A. 1996. Neocortex size and behavioral ecology in primates. *Proceedings of the Royal Society of London B* 263:173–177.

Basabose, A. K. 2004. Fruit availability and chimpanzee party size at Kahuzi montane forest, Democratic Republic of Congo. *Primates* 45:211–219.

Bauer, H. R. 1979. "Agonistic and grooming behavior in the reunion context of Gombe Stream chimpanzees," in *The great apes*. Edited by D. A. Hamburg and E. R. McCown, pp. 395–403. Menlo Park, CA: Benjamin.

Bell, D. 1997. Defining marriage and legitimacy. *Current Anthropology* 38:237–253.

Bernard, H. R., and P. D. Killworth. 1973. On the social structure of an ocean-going research vessel and other important things. *Social Science Research* 2:145–184.

———. 1979. Why are there no social physics? *Journal of the Steward Anthropological Society* 11:33–58.

Bernstein, I. S. 1991. "The correlation between kinship and behaviour in non-human primates," in *Kin recognition*. Edited by P. G. Hepper, pp. 6–29. Cambridge: Cambridge University Press.

Binford, L. 2001. *Constructing frames of reference: An analytical method for archaeological theory building using ethnographic and environmental data sets*. Berkeley: University of California Press.

Blumenschine, R. J., and B. L. Pobiner. 2007. "Zooarchaeology and the ecology of Oldowan hominin carnivory," in *Evolution of the human diet: The known, the unknown, and the unknowable*. Edited by P. S. Ungar, pp. 167–190. Oxford: Oxford University Press.

Boccia, M. L. 1998. "Grooming behavior of primates," in *Comparative psychology: A handbook*. Edited by G. Greenberg and M. M. Haraway, pp. 674–678. London: Routledge.

Boehm, C. 1999. *Hierarchy in the forest: The evolution of egalitarian behavior*. Cambridge, MA: Harvard University Press.

———. 2004. What makes humans economically distinctive? A three-species evolutionary comparison and historical analysis. *Journal of Bioeconomics* 6:109–135.

Boesch, C. 1996. "Social grouping in Taï chimpanzees," in *Great apes societies*. Edited by W. C. McGrew, L. F. Marchant, and T. Nishida, pp. 101–113. Cambridge: Cambridge University Press.

Boesch, C., and H. Boesch. 1983. Optimisation of nut-cracking with natural hammers by wild chimpanzees. *Behaviour* 83:265–286.

Boesch, C., and H. Boesch-Achermann. 2000. *The chimpanzees of the Taï Forest*. Oxford: Oxford University Press.

Boesch, C., P. Marchesi, N. Marchesi, B. Fruth, and F. Joulian. 1993. Is nut cracking in wild chimpanzees a cultural behavior? *Journal of Human Evolution* 26:325–338.

Borries, C., K. Launhardt, C. Epplen, J. T. Epplen, and P. Winkler. 1999. DNA analyses support the hypothesis that infanticide is adaptive in langur monkeys. *Proceedings of the Royal Society of London B* 266:901–904.

Boyd, A. M. 2000. Behavior variability in ring-tailed lemurs (*Lemur catta*), M.A. Thesis, East Carolina University.

Brosnan, S. F., J. B. Silk, J. Henrich, M. C. Mareno, S. P. Lambeth, and S. J. Schapiro. 2009. Chimpanzees (*Pan troglodytes*) do not develop contingent reciprocity in an experimental task. *Animal Cognition* 12:587–597.

Brown, D. 1991. *Human universals*. Philadelphia: Temple University Press.

Burridge, K. O. L. 1959/1960. Siblings in Tangu. *Oceania* 30:127–154.

Carneiro, R. L. 1981. Herbert Spencer as an anthropologist. *Journal of Libertarian Studies* 5:153–210.

Cawthon Lang, K. A. 2005, June 13. "Primate factsheets: Orangutan (Pongo) behavior." <http://pin.primate.wisc.edu/factsheets/entry/orangutan/behav>. Accessed June 21, 2011.

Chapais, B. 1992. "The role of alliances in social inheritance of rank among female primates," in *Coalitions and alliances in humans and other animals*. Edited by A. H. Harcourt and F. B. M. de Waal, pp. 29–59. Oxford: Oxford University Press.

———. 2008. *Primeval kinship: How pair bonding gave birth to human society*. Cambridge, MA: Harvard University Press.

Chapman, C. A., and I. J. Chapman. 2000. "Determinants of group size in primates: The importance of travel costs," in *On the move: How and why animals travel in groups*. Edited by S. Boinski and P. Garber, pp. 24–42. Chicago: University of Chicago Press.

Chapman, C. A., F. J. White, and R. W. Wrangham. 1994. "Party size in chimpanzees and bonobos," in *Chimpanzee cultures*. Edited by R. W. Wrangham, W. C. McGrew, F. B. M. de Waal, and P. G. Heltne, pp. 41–57. Cambridge, MA: Harvard University Press.

Chapman, C. A., R. W. Wrangham, and I. Chapman. 1995. Ecological constraints on group size: An analysis of spider monkey and chimpanzee subgroups. *Behavioral Ecology and Sociobiology* 36:59–70.

Charles-Dominique, P. 1978. "Solitary and gregarious prosimians: Evolution of social structure in primates," in *Recent advances in primatology, Vol. 3*. Edited by D. J. Chivers and K. A. Joysey, pp. 139–149. London: Academic Press.

Cheney, D. L. 1978. Interactions of immature male and female baboons with adult females. *Animal Behaviour* 26:389–408.

———. 1987. "Interactions and relationships between groups," in *Primate societies*. Edited by B. B. Smuts, D. L. Cheney, R. M. Seyfarth, and T. T. Struhsaker. Chicago: University of Chicago Press.

Cheney, D. L., L. R. Moscovice, M. Heesen, R. Mundry, and R. M. Seyfarth. 2010. Contingent cooperation between wild female baboons. *Proceedings of the National Academy of Sciences (USA)* 107:9562–9566.

Cheney, D. L., and R. M. Seyfarth. 2007. *Baboon metaphysics: The evolution of a social mind*. Chicago: University of Chicago Press.

Chism, J., and W. Rogers. 2002. "Grooming and social cohesion in patas monkeys and other guenons," in *The guenons: Diversity and adaptation in African monkeys*. Edited by M. E. Glenn and M. Cords, pp. 233–244. New York: Kluwer Academic.

Collings, P., G. Wenzel, and R. G. Condon. 1998. Modern food sharing networks and community integration in the Central Canadian Arctic. *Arctic* 51:301–314.

Coolidge, F. L., and T. Wynn. 2009. *The rise of Homo sapiens: The evolution of modern thinking*. Oxford: Wiley-Blackwell.

Corballis, M. C. 2007. "The evolution of consciousness," in *The Cambridge handbook of consciousness*. Edited by P. D. Zelazo, M. Moscovitch, and E. Thompson, pp. 571–614. Cambridge: Cambridge University Press.

Cords, M. 2002. Friendship among adult female blue monkeys (*Cercopithecus mitis*). *Behaviour* 139:291–314.

Curr, E. M. 1886 [2005]. *The Australian race: Its origin, languages, customs, place of landing in Australia and the routes by which it spread itself over that continent*. Vol. I. Boston: Adamant Media Corporation.

Damas, D. 1972. Central Eskimo systems of food sharing. *Ethnology* 11:220–240.

de Waal, F. B. M. 1982. *Chimpanzee politics*. London: Jonathan Cape.

———. 1989. "Dominance 'style' and primate social organization," in *Comparative socioecology: The behavioural ecology of humans and other mammals*. Edited by V. Standen and R. A. Foley, pp. 243–263. London: Blackwell.

———. 1998. *Chimpanzee politics: Power and sex among apes*, Revised edition. Baltimore: Johns Hopkins Press.

———. 2000. Primates—a natural heritage of conflict resolution. *Science* 289:586–590.

de Waal, F. B. M., and F. Aureli. 1996. "Consolation, reconciliation, and a possible cognitive difference between macaques and chimpanzees," in *Reaching into thought: The minds of the great apes*. Edited by A. E. Russon, K. A. Bard, and S. T. Parker, pp. 80–110. Cambridge: Cambridge University Press.

de Waal, F. B. M., and P. L. Tyack. 2003. "Preface," in *Animal social complexity—intelligence, culture, and individualized societies*. Edited by F. B. M. de Waal and P. L. Tyack, pp. ix–xiv. Cambridge, MA: Harvard University Press.

Di Fiore, A., and D. Rendall. 1994. Evolution of social organization: A reappraisal for primates by using phylogenetic methods. *Proceedings of the National Academy of Sciences (USA)* 91:9941–9945.

Doran, D. M., W. L. Jungers, Y. Sugiyama, J. G. Fleagle, and C. P. Heesy. 2002. "Multivariate and phylogenetic approaches to understanding chimpanzee and bonobo behavioral diversity," in *Behavioural diversity in chimpanzees and bonobos*. Edited by C. Boesch, G. Hohmann, and L. F. Marchant, pp. 14–34. Cambridge: Cambridge University Press.

Dousset, L. 2005. Structure and substance: Combining 'classic' and 'modern' kinship studies in the Australian Western Desert. *The Australian Journal of Anthropology* 16:18–30.

Dubreuil, B. 2010. *Human evolution and the origins of hierarchies: The state of nature.* Cambridge: Cambridge University Press.

Dunbar, R. I. M. 1988. *Primate social systems.* London: Croom Helm.

———. 1992. A hidden constraint on the behavioural ecology of baboons. *Behavioral Ecology and Sociobiology* 31:35–49.

———. 1993. Co-evolution of neocortex size, group size and language in humans. *Behavioral and Brain Sciences* 16:681–735.

———. 1998. The social brain hypothesis. *Evolutionary Anthropology* 6:178–190.

Durham, W. 1991. *Coevolution: Genes, culture, and human diversity.* Stanford, CA: Stanford University Press.

Epstein, H. T. 2002. Evolution of the reasoning brain. *Behavioral and Brain Sciences* 25:408–409.

Fairbanks, L. A. 2000. "Maternal investment throughout the life span in Old World monkeys," in *Old world monkeys*. Edited by P. F. Whitehead and C. J. Jolly, pp. 341–367. Cambridge: Cambridge University Press.

Falk, D. 2004. Prelinguistic evolution in early hominins: Whence motherese? *Behavioral and Brain Sciences* 27:491–541.

Fararo, T. J. 1997. Reflections on mathematical sociology. *Sociological Forum* 12: 73–101.

Fedurek, P., and R. I. M. Dunbar. 2009. What does mutual grooming tell us about why chimpanzees groom? *Ethology* 115:566–575.

Fernandez-Duque, E., C. R. Valeggia, and S. P. Mendoza. 2009. The biology of paternal care in human and nonhuman primates. *Annual Review of Anthropology* 38:115–130.

Focquaert, F., J. Braeckman, and S. M. Platek. 2008. An evolutionary cognitive neuroscience perspective on human self-awareness and Theory of Mind. *Philosophical Psychology* 21:47–68.

Fortes, M. 1969. *Kinship and the social order: The legacy of Lewis Henry Morgan.* Chicago: Aldine Publishing Company.

Foster, M. W., I. C. Gilby, C. M. Murray, A. Johnson, E. Wroblewski, and A. Pusey. 2008. Alpha male chimpanzee grooming patterns: Implications for dominance "style." *American Journal of Primatology* 71:136–144.

Furuichi, T. 1997. Agonistic interactions and matrifocal dominance rank of wild bonobos (*Pan paniscus*) at Wamba. *International Journal of Primatology* 18:855—875.

———. 2009. Factors underlying party size differences between chimpanzees and bonobos: A review and hypotheses for future study. *Primates* 50:197–209.

Furuichi, T., and H. Ihobe. 1994. Variation in male relationships in bonobos and chimpanzees. *Behaviour* 130:211–228.

Gadsby, P. 2004. The Inuit paradox. *Discover* 25:48–54.

Galdikas, B. M. F. 1984. "Adult female sociality among wild orangutans at Tanjung Puting Reserve," in *Female primates: Studies by women primatologists.* Edited by M. F. Small, pp. 217–235. New York: Alan R. Liss.

Gomes, C. M., and C. Boesch. 2009. Wild chimpanzees exchange meat for sex on a long-term basis. *PLoS ONE* 44:e5116.

Goodale, J. C. 1971. *Tiwi wives: A study of the women of Melville Island, North Australia.* Seattle: University of Washington Press.

Goodall, J. 1986. *The chimpanzees of Gombe.* Cambridge, MA: Belknap Press.

Gough, K. 1959. The Nayar and the definition of marriage. *Journal of the Royal Anthropological Institute* 89:23–34.

Gouzoules, S., and H. Gouzoules. 1987. "Kinship," in *Primate societies.* Edited by B. B. Smuts, D. L. Cheney, R. M. Seyfarth, R. W. Wrangham, and T. T. Struhsaker, pp. 299–305. Chicago: University of Chicago Press.

Harcourt, A. H., and K. J. Stewart. 2007. *Gorilla society: Conflict, compromise, and cooperation between the sexes.* Chicago: University of Chicago Press.

Harcourt-Smith, W. E. H., and L. C. Aiello. 2004. Fossils, feet and the evolution of human bipedal locomotion. *Journal of Anatomy* 204:403–416.

Hashimoto, C., T. Furuichi, and Y. Tashiro. 2001. What factors affect the size of chimpanzee parties in the Kalinzu Forest, Uganda? Examination of fruit abundance and number of estrous females. *International Journal of Primatology* 22:947–959.

Hassan, F. A. 1978. Demographic archaeology. *Advances in Archaeological Method and Theory* 1:49–103.

Henzi, S. P., and L. Barrett. 1999. The value of grooming to female primates. *Primates* 40:47–59.

Henzi, S. P., J. E. Lycett, and S. E. Piper. 1997. Fission and troop size in a mountain baboon population. *Animal Behaviour* 53:525–535.

Hill, D. A. 1999. Effects of provisioning on the social behaviour of Japanese and rhesus macaques: Implications for socioecology. *Primates* 40:187–198.

Hill, K., R. S. Walker, M. Božičević, J. Eder, T. Headland, B. S. Hewlett, A. M. Hurtado, F. Marlowe, P. Wiessner, and B. Wood. 2011. Co-residence patterns in hunter-gatherer societies show unique human social structure. *Science* 331:1286–1289.

Hill, R. A., L. Barrett, D. Gaynor, T. Weingrill, P. Dixon, H. Payne, and S. P. Henzi. 2003. Day length, latitude and behavioural (in)flexibility in Baboons (*Papio cynocephalus ursinus*). *Behavioral Ecology and Sociobiology* 53:278–286.

Hohmann, G. 2001. Association and social interactions between strangers and residents in bonobos (*Pan paniscus*). *Primates* 42:91–99.

———. 2009. "The diets of non-human primates: Frugivory, food processing, and food sharing," in *The evolution of hominin diets: Integrating approaches to the study of Paleolithic subsistence*. Edited by J.-J. Hublin and M. P. Richards, pp. 1–14. Berlin: Springer.

Hohmann, G., and B. Fruth. 2002. "Dynamics in social organization of bonobos (*Pan paniscus*)," in *Behavioural diversity in chimpanzees and bonobos*. Edited by C. Boesch, G. Hohmann, and L. F. Marchant, pp. 138–150. Cambridge: Cambridge University Press.

Homewood, K. M. 1975. Can the Tana Mangabey survive? *Oryx* 13:53–59.

———. 1978. Feeding strategy of the Tana Mangabey (*Cercocebus galeritus*) (Mammalia: Primates). *Journal of Zoology, London* 186:375–391.

Hughes, S. S. 1998. Getting to the point: Evolutionary change in prehistoric weaponry. *Journal of Archaeological Method and Theory* 5:345–407.

Ibrahim, T. 1997. *An-Nawawi's Forty Hadith*. San Francisco: Ignatius Press.

Ingmanson, and E. J. 1996. "Tool-using behavior in wild *Pan paniscus*: Social and ecological considerations," in *Reaching into thought: The minds of the great apes*. Edited by A. E. Russon, K. A. Bard, and S. T. Parker, pp. 190–210. Cambridge: Cambridge University Press.

Inogwabini, B. I., and I. Omari. 2005. A landscape-wide distribution of *Pan paniscus* in the Salonga National Park, Democratic Republic of Congo. *Endangered Species Update* 22:116–123.

Izar, P. 2004. Female social relationships of *Cebus apella nigritus* in a southeastern Atlantic forest: An analysis through ecological models of primate social evolution. *Behaviour* 141:71–99.

Janson, C. H. 1988. Intra-specific food competition and primate social structure: A synthesis. *Behaviour* 105:1–17.

Jolly, A. 1998. Lemur social structure. *Folia Primatologica* 69(suppl.1):1–13.

Jolly, C. J. 1966. *Lemur behavior*. Chicago: University of Chicago Press.

Kaplan, H., K. Hill, J. Lancaster, and A. M. Hurtado. 2000. A theory of human life history evolution: Diet, intelligence, and longevity. *Evolutionary Anthropology* 9:156–185.

Kappeler, P. M. 1993a. Reconciliation and post-conflict behavior in ringtailed (*Lemur catta*) and redfronted (*Eulemur fulvus rufus*) lemurs. *Animal Behaviour* 45:901–915.

———. 1993b. Variation in social structure: The effects of sex and kinship on social interactions in three lemur species. *Ethology* 93:125–145.

Kappeler, P. M., and C. P. van Schaik. 1992. Methodological and evolutionary aspects of reconciliation among primates. *Ethology* 92:51–69.

———. 2002. Evolution of primate social systems. *International Journal of Primatology* 23:707–740.

Kawamura, S. 1965. "Matriarchal social ranks in the Minoo-B troop: A study of the rank system of Japanese monkeys," in *Japanese monkeys*. Edited by K. Imanishi and S. A. Altmann, pp. 105–112. Chicago: Altmann.

Kelly, R. L. 1995. *The foraging spectrum: Diversity in hunter-gatherer lifeways.* Washington, DC: Smithsonian Institution Press.

Klein, R. G. 1999. *The human career: Human biological and cultural origins*, 2nd edition. Chicago: University of Chicago Press.

Knott, C., L. Beaudrot, T. Snaith, S. White, H. Tschauner, and G. Planansky. 2008. Female-female competition in Bornean orangutans. *International Journal of Primatology* 29:975–997.

Koenig, A., and C. Borries. 2001. Socioecology of Hanuman Langurs: The story of their success. *Evolutionary Anthropology* 10:122–137.

Konner, M. J. 1972. "Aspects of the development ethology of a foraging people," in *Ethological studies of child behaviour*. Edited by N. Blurton Jones, pp. 285–304. Cambridge: Cambridge University Press.

Kudo, H., and R. I. M. Dunbar. 2001. Neocortex size and social network size in primates. *Animal Behaviour* 62:711–722.

Lane, D., D. Pumain, and S. van der Leeuw. 2009. "Introduction," in *Complexity perspectives in innovation and social change*. Edited by D. Lane, D. Pumain, S. van der Leeuw, and G. West, pp. 1–7. Berlin: Springer.

Lane, D., R. M. Maxfield, D. Read, and S. van der Leeuw. 2009. "From population to organization thinking," in *Complexity perspectives on innovation and social change*. Edited by D. Lane, D. Pumain, S. van der Leeuw, and G. West, pp. 43–84. Berlin: Springer.

Langergraber, K. E., C. Boesch, E. Inoue, M. Inoue-Murayama, J. C. Mitani, T. Nishida, A. Pusey, V. Reynolds, G. Schubert, R. W. Wrangham, E. Wroblewski, and L. Vigilant. 2011. Genetic and 'cultural' similarity in wild chimpanzees. *Proceedings of the Royal Society of London B-Biological Sciences* 278:408–416.

Langergraber, K. E., J. C. Mitani, and L. Vigilant. 2007. The limited impact of kinship on cooperation in wild chimpanzees. *Proceedings of the National Academy of Sciences (USA)* 104:7786–7790.

———. 2009. Kinship and social bonds in female chimpanzees (*Pan troglodytes*). *American Journal of Primatology* 71:840–851.

Leaf, M. 2006. Experimental analysis of kinship. *Ethnology* 45:305–330.

Leaf, M., and D. Read. Forthcoming. *The conceptual foundation of human society and thought: Anthropology on a new plane.* Lanham, MD: Lexington Books.

Lee, R. B. 1979. *The !Kung San: Men, women, and work in a foraging society.* Cambridge: Cambridge University Press.

Lehmann, J., and C. Boesch. 2004. To fission or to fusion: Effects of community size on wild chimpanzee (*Pan troglodytes verus*) social organization. *Behavioral Ecology and Sociobiology* 56:207–216.

———. 2008. Sexual differences in chimpanzee sociality. *International Journal of Primatology* 29:65–81.

Lehmann, J., A. H. Korstjens, and R. I. M. Dunbar. 2007a. Group size, grooming and social cohesion in primates. *Animal Behaviour* 74:1617–1629.

———. 2007b. Fission–fusion social systems as a strategy for coping with ecological constraints: A primate case. *Evolutionary Ecology* 21:613–634.

———. 2008. Time and distribution: A model of ape biogeography. *Ethology Ecology & Evolution* 20:337–359.

Lévi-Strauss, C. 1969[1949]. *Elementary structures of kinship.* Oxford: Mowden and Albray.

———. 1963. *The raw and the cooked: Introduction to a science of mythology: 1.* New York: Penguin Books.

Levinson, S. 2002. Matrilineal clans and kin terms on Rossel Island. *Anthropological Linguistics* 48:1–43.

Livingston, F. 1958. Anthropological implications of sickle cell gene distribution in West Africa. *American Anthropologist* 60:533–562.

Lovejoy, O. 2005. The natural history of human gait and posture Part 1. Spine and pelvis. *Gait and Posture* 21:95–112.

Lycett, S. J., M. Collard, and W. C. McGrew. 2010. Are behavioral differences among wild chimpanzee communities genetic or cultural? An assessment using tool-use data and phylogenetic methods. *American Journal of Physical Anthropology* 142:461–467.

MacLeod, C. E. 2004. "What's in a brain? The question of a distinctive brain anatomy in great apes," in *The evolution of thought: Evolutionary origins of great ape intelligence.* Edited by A. E. Russon and D. R. Begun, pp. 105–121. Cambridge: Cambridge University Press.

Marlowe, F. 2004. Marital residence among foragers. *Current Anthropology* 45:277–284.

Martin, R. D. 1981. Relative brain size and basal metabolic rate in terrestrial vertebrates. *Nature* 293:57–60.

Maryanski, A. 1987. African ape social structure: Is there strength in weak ties? *Social Networks* 9:191–215.

Maryanski, A., and J. H. Turner. 1992. *The social cage: Human nature and the evolution of society.* Stanford, CA: Stanford University Press.

Matsumoto-Oda, A., K. Hosaka, M. A. Huffman, and K. Kawanaka. 1998. Factors affecting party size in chimpanzees of the Mahale Mountains. *International Journal of Primatology* 19:999–1011.

Matsuzawa, T. 1994. "Field experiments on use of stone tools by chimpanzees in the wild," in *Chimpanzee cultures.* Edited by R. W. Wrangham, W. C. McGrew, F. B. M. de Waal, and P. G. Heltne, pp. 351–370. Cambridge, MA: Harvard University Press.

Maxwell, J. 1994. "Biology and social relationship in the kin terminology of an Inuit community," in *North American Indian anthropology: Essays on society and culture.* Edited by R. J. DeMalie and A. Ortiz, pp. 25–48. Norman: University of Oklahoma Press.

McGrew, W. C. 2003a. "Ten dispatches from the chimpanzee culture wars," in *Animal social behavior.* Edited by F. B. M. de Waal and P. L. Tyack, pp. 419–439. Cambridge, MA: Harvard University Press.

———. 2003b. *The cultured chimpanzee.* Cambridge: Cambridge University Press.

McGrew, W. C., and C. E. G. Tutin. 1978. Evidence for a social custom in wild chimpanzees? *Man (N.S.)* 13:234–251.

Melis, A. P., B. Hare, and M. Tomasello. 2008. Do chimpanzees reciprocate received favours? *Animal Behaviour* 76:951–962.

Mellars, P. 1973. "The character of the Middle-Upper Palaeolithic transition in south-west France," in *The explanation of culture change.* Edited by C. Renfrew, pp. 255–276. London: Duckworth.

———. 1982. On the Middle/Upper Palaeolithic transition: A reply to White. *Current Anthropology* 23:238–240.

Menard, N., and D. Vallet. 1997. Behavioral responses of Barbary macaques (*Macaca sylvanus*) to variations in environmental conditions in Algeria. *American Journal of Primatology* 43:285–304.

Miller, G. A. 1956. The magical number seven, plus or minus two: Some limits on our capacity for processing information. *Psychological Review* 63:81–97.

Milton, K. 1984. Habitat, diet, and activity patterns of free-ranging woolly spider monkeys (*Brachyteles arachnoides* E. Geoffroy 1806). *International Journal of Primatology* 5:491–514.

Missakian, E. A. 1974. Mother-offspring grooming relations in Rhesus monkeys. *Archives of Sexual Behavior* 3:135–141.

Mitani, J. C. 2006a. Demographic influences on the behavior of chimpanzees. *Primates* 47:6–13.

———. 2006b. "Reciprocal exchanges in chimpanzees and other primates," in *Cooperation in primates and humans*. Edited by P. M. Kappeler and C. P. van Schaik, pp. 107–119. Berlin: Springer.

———. 2009a. Cooperation and competition in chimpanzees: Current understanding and future challenges. *Evolutionary Anthropology* 18:215–227.

———. 2009b. Male chimpanzees form enduring and equitable social bonds. *Animal Behaviour* 77:633–640.

Mitani, J. C., and S. J. Amsler. 2003. Social and spatial aspects of male subgrouping in a community of wild chimpanzees. *Behaviour* 140:869–884.

Mitani, J. C., D. A. Merriwether, and C. Zhang. 2000. Male affiliation, cooperation and kinship in wild chimpanzees *Animal Behaviour* 59:885–893.

Mitani, J. C., D. P. Watts, and S. J. Amsler. 2010. Lethal intergroup aggression leads to territorial expansion in wild chimpanzees. *Current Biology* 20:R507–R508.

Mitani, J. C., D. P. Watts, and J. S. Lwanga. 2002. "Ecological and social correlates of chimpanzee party size and composition," in *Behavioral diversity in chimpanzees and bonobos*. Edited by C. Boesch, G. Hohmann, and L. F. Marchant, pp. 102–111. Cambridge: Cambridge University Press.

Mitani, J. C., D. P. Watts, and M. N. Muller. 2002. Recent developments in the study of wild chimpanzee behavior. *Evolutionary Anthropology* 11:9–25.

Moore, J. 1992. Dispersal, nepotism, and primate social behavior. *International Journal of Primatology* 13:361–378.

Mulavwa, M., T. Furuichi, K. Yangozene, M. Yamba-Yamba, B. Motema-Salo, G. Idani, H. Ihobe, C. Hashimoto, Y. Tashiro, and N. Mwanza. 2008. "Seasonal changes in fruit production and party size of bonobos at Wamba," in *The bonobos: Behavior, ecology, and conservation*. Edited by T. Furuichi and J. M. Thompson, pp. 120–134. New York: Springer.

Muller, M. N. 2002. "Agonistic relations among Kanyawara chimpanzees," in *Behavioural diversity in chimpanzees and bonobos*. Edited by C. Boesch, G. Hohmann, and L. F. Marchant, pp. 112–123. Cambridge: Cambridge University Press.

Muller, M. N., and J. C. Mitani. 2005. Conflict and cooperation in wild chimpanzees. *Advances in the Study of Behavior* 35:275–331.

Nakamichi, M. 1989. Sex differences in social development during the first 4 years in a free-ranging group of Japanese monkeys, *Macaca fuscata. Animal Behaviour* 38:737–748.

Nakamura, M. 2000. Is human conversation more efficient than chimpanzee grooming?: Comparison of clique sizes. *Human Nature* 11:281–297.

———. 2003. 'Gatherings' of social grooming among wild chimpanzees: Implications for evolution of sociality. *Journal of Human Evolution* 44:59–72.

Nakamura, M., W. C. McGrew, and L. F. Marchant. 2000. Social scratch: Another custom in wild chimpanzees? *Primates* 41:237–248.

Neuberger, J. 2005. Embryos and ensoulment: When does life begin? *The Lancet* 365:836–838.

Newton, P. N., and R. I. M. Dunbar. 1994. "Colobine monkey society," in *Colobine monkeys: Their ecology, behaviour and evolution*. Edited by A. G. Davies and J. F. Oates, pp. 311–346. Chicago: University of Chicago Press.

Newton-Fisher, N. E. 1997. Tactical behaviour and decision making in wild chimpanzees. PhD diss., Cambridge University.

———. 2002. "Relationships of male chimpanzees in the Budongo Forest, Uganda," in *Behavioural diversity in chimpanzees and bonobos*. Edited by C. Boesch, G. Hohmann, and M. L. Frances, pp. 125–137. Cambridge: Cambridge University Press.

Newton-Fisher, N. E., V. Reynolds, and A. J. Plumptre. 2000. Food supply and chimpanzee (*Pan troglodytes schweinfurthii*) party size in the Budongo Forest Reserve, Uganda. *International Journal of Primatology* 21:613–628.

Nishida, T. 1979. "The social structure of chimpanzees of the Mahale Mountains," in *The great apes*. Edited by D. A. Hamburg and E. R. McCown, pp. 72–121. Menlo Park, CA: Benjamin/Cummings.

———. 1979. "The social structure of chimpanzees of the Mahale Mountains," in *The great apes*. Edited by D. A. Hamburg and E. R. McCown, pp. 72–121. Menlo Park, CA: Benjamin/Cummings.

Nishida, T., M. Hiraiwa-Hasegawa, T. Hasegawa, and Y. Takahata. 1985. Group extinction and female transfer in wild chimpanzees in the Mahale Mountains National Park, Tanzania. *Zeitschrift für Tierpsychologie* 67:281–301.

Nishida, T., and K. Hosaka. 1996. "Coalition strategies among adult male chimpanzees of the Mahale Mountains, Tanzania," in *Great ape societies*. Edited by L. F. Marchant and T. Nishida, pp. 114–134. Cambridge: Cambridge University Press.

Nishida, T., J. C. Mitani, and D. P. Watts. 2003. Variable grooming in wild chimpanzees. *Folia Primatologica* 75:31–36.

Nishida, T., C. N., M. Hamai, T. Hasegawa, M. Hiraiwa-Hasegawa, K. Hosaka, K. Hunt, N. Itoh, K. Kawanaka, A. Matsumoto-Oda, J. C. Mitani, M. Nakamura, K. Norikoshi, T. Sakamaki, L. Turner, S. Uehara, and K. Zamma. 2003. Demography, female life history, and reproductive profiles among the chimpanzees of Mahale. *American Journal of Primatology* 59:99–121.

Nissen, H. W. 1956. Individuality in the behavior of chimpanzees. *American Anthropologist* 58:407–413.

Okalik, P. 2001. "Brisbane Dialogue: What does indigenous government mean?" Paper presented at Customs House, Brisbane Institute, Brisbane, Australia. August 13.

Ottenheimer, M. 1996. *Forbidden relatives: The American myth of cousin marriage*. Urbana: University of Illinois Press.

Palmer, C. T., and L. B. Steadman. 1997. Human kinship as a descendant-leaving strategy: A solution to an evolutionary puzzle. *Journal of Social and Evolutionary Systems* 20:39–51.

Parker, S. T. 2004. "The cognitive complexity of social organization and socialization in wild baboons and chimpanzees: Guided participation, socializing interactions, and event representation," in *The evolution of thought: Evolutionary origins of great ape intelligence*. Edited by A. E. Russon and D. R. Begun, pp. 45–60. Cambridge: Cambridge University Press.

Parsons, T. 1964. *The social system*. New York: Free Press of Glencoe.

Passingham, R. E. 1975. Changes in the size and organization of the brain in man and his ancestors. *Brain, Behavior and Evolution* 11:73–90.

Patterson, N., D. C. Petersen, R. E. van der Ross, H. Sudoyo, R. Glashoff, S. Marzuki, D. Reich, and V. M. Hayes. 2010. Genetic structure of a unique admixed population: Implications for medical research. *Human Molecular Genetics* 19:411–419.

Penn, D. C., K. J. Holyoak, and D. J. Povinelli. 2008. Darwin's mistake: Explaining the discontinuity between human and nonhuman minds. *Behavioral and Brain Sciences* 31:109–130.

Poirier, F. E. 1969. The Nilgiri langur (*Presbytis johnii*) troop: Its composition, structure, function and change. *Folia Primatologica* 10:20–47.

———. 1970. "The Nilgiri langur (*Presbytis johnii*) of South India," in *Primate behavior—Developments in field and laboratory research*. Edited by L. A. Rosenblum, pp. 254–383. New York: Academic Press.

Pokempner, A. A. 2009. Fission-fusion and foraging: Sex differences in the behavioral ecology of chimpanzees (*Pan troglodytes schweinfurthii*). PhD diss., Stony Brook University.

Potts, R. 2003. "Early human predation," in *Predator-prey interactions in the fossil record*. Edited by P. H. Kelley, M. Kowalewski, and T. A. Hansen, pp. 359–377. New York: Kluwer Academic/Plenum.

Povinelli, D. J., and J. Vonk. 2004. We don't need a microscope to explore the chimpanzee's mind. *Mind and Language* 19:1–28.

Pruetz, J. D., and P. Bertolani. 2007. Savanna chimpanzees, *Pan troglodytes verus*, hunt with tools. *Current Biology* 17:412–417.

Radcliffe-Brown, A. R. 1913. Three tribes of western Australia. *Journal of the Royal Anthropological Institute (N.S.)* 43:143–194.

———. 1950. "Introduction," in *African systems of kinship and marriage*. Edited by A. R. Radcliffe-Brown and D. Forde, pp. 1–85. London: Oxford University Press.

Raichlen, D. A., A. D. Gordon, W. E. H. Harcourt-Smith, A. D. Foster, and W. R. Haas Jr. 2010. Laetoli footprints preserve earliest direct evidence of human-like bipedal biomechanics. *PLoS ONE* 5(3):e9769.

Read, D. 1987. Foraging society organization: A simple model of a complex transition. *European Journal of Operational Research* 30:230–236.

———. 1998. Kinship based demographic simulation of societal processes. *Journal of Artificial Societies and Social Simulation*, vol. 1. no. 1. Retrieved from http://jasss.soc.surrey.ac.uk/1/1/1.html.

———. 2001. "What is kinship?" in *The cultural analysis of kinship: The legacy of David Schneider and its implications for anthropological relativism*. Edited by R. Feinberg and M. Ottenheimer, pp. 78–117. Urbana: University of Illinois Press.

———. 2005. "Quantitative analysis, anthropology," in *Encyclopedia of social measurement, Vol. 3*. Edited by K. Kempf-Leonard, pp. 237–246. Amsterdam: Elsevier Press.

———. 2006. Some observations on resilience and robustness in human systems. *Cybernetics and Systems (Special Issue)* 36:773–802.

———. 2007. Kinship theory: A paradigm shift. *Ethnology* 46:329–364.

———. 2008a. "A formal explanation of formal explanation," in *Structure and Dynamics* 3(2). Retrieved from: http://escholarship.org/uc/item/91z973j6.

———. 2008b. An interaction model for resource implement complexity based on risk and number of annual moves. *American Antiquity* 73:599–625.

———. 2008c. Working memory: A cognitive limit to non-human primate recursive thinking prior to hominid evolution. *Evolutionary Psychology* 6:603–643.

———. 2010a. Agent-based and multi-agent simulations: Coming of age or in search of an identity? *Computational and Mathematical Organization Theory* 16:329–347.

———. 2010b. "From experiential-based to relational-based forms of social organization: A major transition in the evolution of *Homo sapiens*," in *Social brain, distributed mind*. Edited by R. Dunbar, C. Gamble, and J. Gowlett, pp. 199–230. Oxford: Oxford University Press.

Read, D., and C. Behrens. 1990. KAES: An expert system for the algebraic analysis of kinship terminologies. *Journal of Quantitative Anthropology* 2:353–393.

Read, D., M. D. Fischer, and F. K. L. Chit Hlaing. Forthcoming. "The cultural grounding of kinship: A paradigm shift," in *La parenté en débat: Penser la parenté aujourd'hui*. Edited by L. Barry, K. Hamberger, and M. Houseman. Paris: Editions de la Maison des Sciences de l'Homme.

Read, D., D. Lane, and S. van der Leeuw. 2009. "The innovation innovation," in *Complexity perspectives in innovation and social change*. Edited by D. Lane, D. Pumain, S. van der Leeuw, and G. West, pp. 43–84. Berlin: Springer.

Read, D., and S. LeBlanc. 2003. Population growth, carrying capacity, and conflict. *Current Anthropology* 44:59–85.

Read, D., and S. van der Leeuw. 2008. Biology is only part of the story. *Philosophical Transactions of the Royal Society B* 363:1959–1968.

Ridington, R., and J. Ridington. 2006. *When you sing it now, just like new: First Nations poetics, voices and representations.* Lincoln: University of Nebraska Press.

Rightmire, G. P. 2004. Brain size and encephalization in Early to Mid-Pleistocene *Homo. American Journal of Physical Anthropology* 124:109–123.

Rilling, J. K. 2006. Human and nonhuman primate brains: Are they allometrically scaled versions of the same design? *Evolutionary Anthropology* 15:65–77.

Rilling, J. K., J. Scholz, T. M. Preuss, M. F. Glasser, B. K. Errangi, and T. E. Behrens. 2011. Differences between chimpanzees and bonobos in neural systems supporting social cognition. *Social Cognitive and Affective Neuroscience.* ePub ahead of print. http://scan.oxfordjournals.org/content/early/2011/04/04/scan.nsr017.full.pdf+html. Accessed June 28, 2011.

Robson, S. L., and B. Wood. 2008. Hominin life history: Reconstruction and evolution. *Journal of Anatomy* 212:394–425.

Rogers, A. R., D. Iltis, and S. Wooding. 2004. Genetic variation at the MC1R locus and the time since loss of human body hair. *Current Anthropology* 45:105–108.

Roney, J. R., and D. Maestripieri. 2003. "Social development and affiliation," in *Primate psychology.* Edited by D. Maestripieri, pp. 171–204. Cambridge, MA: Harvard University Press.

Ruff, C. B., E. Trinkhaus, and T. W. Holliday. 1997. Body mass and encephalization in Pleistocene *Homo. Nature* 387:173–176.

Russon, A. E. 2004. "Great ape cognitive systems," in *The evolution of thought: Evolutionary origins of great ape intelligence.* Edited by A. E. Russon and D. R. Begun, pp. 76–100. Cambridge: Cambridge University Press.

Russon, A. E., and K. A. Bard. 1996. "Exploring the minds of the great apes: Issues and controversies," in *Reaching into thought: The minds of the great apes.* Edited by A. E. Russon, K. A. Bard, and S. T. Parker, pp. 1–22. Cambridge: Cambridge University Press.

Sambrook, T. D., A. Whiten, and S. C. Strum. 1995. Priority of access and grooming patterns of females in a large and a small group of olive baboons. *Animal Behaviour* 50:1667–1682.

Sayers, K. A. 2008. Optimal foraging on the roof of the world: A field study of Himalayan langurs. PhD diss., Kent State University.

Schieffelin, E. L. 1976[2005]. *The sorrow of the lonely and the burning of the dancers.* New York: Palgrave Macmillan.

Schino, G. 2001. Grooming competition and social rank among female primates: A meta-analysis. *Animal Behaviour* 62:265–271.

———. 2007. Grooming and agonistic support: A meta-analysis of primate reciprocal altruism. *Behavioral Ecology* 18:115–120.

Schino, G., and F. Aureli. 2008. Grooming reciprocation among female primates: A meta-analysis. *Biology Letters* 4:9–11.

———. 2009. Reciprocal altruism in primates: Partner choice, cognition, and emotions. *Advances in the Study of Behavior* 39:45–69.

Service, E. R. 1962. *Primitive social organization: An evolutionary perspective.* New York: Random House.

Seyfarth, R. M. 1977. A model of social grooming among adult female monkeys. *Journal of Theoretical Biology* 65:671–698.

———. 1980. The distribution of grooming and related behaviours among adult female vervet monkeys. *Animal Behaviour* 28:798–813.

Seyfarth, R. M., and D. L. Cheney. 2000. Social awareness in monkeys. *American Zoologist* 40:902–909.

Shih, C-k. 2010. *Quest for harmony: The Moso traditions of sexual union and family life.* Stanford, CA: Stanford University Press.

Shuster, K. 1992. A Halachic overview of abortion. *Suffolk University Law Review* 26:641–651.

Silk, J. 2002. Kin selection in primate groups. *International Journal of Primatology* 23:849–875.

Silk, J., S. C. Alberts, and J. Altmann. 2003. Social bonds of female baboons enhance infant survival. *Science* 302:1231–1234.

Smuts, B. B., and D. J. Gubernick. 1992. "Male-infant relationships in nonhuman primates: Paternal investment or mating effort? " in *Father-child relations: Cultural and biosocial contexts.* Edited by B. S. Hewlett, pp. 1–30. New Brunswick, NJ: Transaction.

Snodgrass, J. J., W. R. Leonard, and M. L. Robertson. 2009. "The energetics of encephalization in early hominids," in *The evolution of hominin diets: Integrating approaches to the study of Paleolithic subsistence.* Edited by J-J. Hublin and M. P. Richards, pp. 15–30. Berlin: Springer.

Sober, E., and D. S. Wilson. 1998. *Unto others: The evolution and psychology of unselfish behavior.* Cambridge, MA: Harvard University Press.

Sockol, M. D., D. A. Raichlen, and H. Pontzer. 2007. Chimpanzee locomotor energetics and the origin of human bipedalism. *Proceedings of the National Academy of Sciences (USA)* 104:12265–12269.

Spencer, H. 1900. *First principles*, 6th edition. New York: D. Appleton and Company.

———. 1910. *The principles of sociology*, 3rd edition. Vol. 1. New York: D. Appleton and Company.

———. 1851. *Social statics.* London: Chapman.

Spinozzi, G., F. Natale, J. Langer, and K. E. Brakke. 1999. Spontaneous class grouping behavior by bonobos (*Pan paniscus*) and common chimpanzees (*P. troglodytes*). *Animal Cognition* 2:157–170.

Stanford, C. B. 1991. *The capped langur in Bangladesh: Behavioral ecology and reproductive tactics*. Basel, Switzerland: Karger.

Sterck, E. H. M., D. P. Watts, and C. P. van Schaik. 1997. The evolution of female social relationships in nonhuman primates. *Behavioral Ecology and Sociobiology* 41:291–309.

Steward, J. H. 1933. Ethnography of the Owens Valley Paiute. *University of California Publications in American Archaeology and Ethnology* 33:233–250.

———. 1936. "The economic and social basis of primitive bands," in *Essays in anthropology in honor of Alfred Louis Kroeber*. Edited by R. H. Lowie, pp. 331–350. Berkeley: University of California Press.

———. 1955. *Theory of culture change: The methodology of multilinear evolution*. Urbana: University of Illinois Press.

Stiles, D. 2001. Hunter-gatherer studies: The importance of context. *African Study Monographs* Suppl.26:41–65.

Strier, K. B. 2000. *Primate behavioral ecology*. Boston: Allyn and Bacom.

Strum, S. C., and B. Latour. 1987. Redefining the social link: From baboons to humans. *Social Science Information* 26:783–802.

Supriatna, J., B. O. Manullang, and E. Soekara. 1986. Group composition, home range, and diet of the maroon leaf monkey (*Presbytis rubicunda*) at Tanjung Puting Reserve, Central Kalimantan, Indonesia. *Primates* 27:185–190.

Surbeck, M., R. Mundry, and G. Hohmann. 2010. Mothers matter! Maternal support, dominance status and mating success in male bonobos (*Pan paniscus*). *Proceedings of the Royal Society of London B*:1–9.

Sussman, R. W. 2003. *Primate ecology and social structure*. Boston: Pearson Custom.

Szalay, F. S., and R. K. Costello. 1991. Evolution of permanent estrus displays in hominids. *Journal of Human Evolution* 20:439–464.

Tonkinson, R. 1991[1978]. *The Mardu Aborigines: Living the dream in Australia's desert. Case Studies in Cultural Anthropology*. New York: Holt, Rinehart and Winston.

Torrence, R. 1989. "Re-tooling: Towards a behavioral theory of stone tools," in *Time, energy and stone tools*. Edited by R. Torrence, pp. 57–66. Cambridge: Cambridge University Press.

———. 2001. "Hunter-gatherer technology: Macro- and microscale approaches," in *Hunter-gatherers: An interdisciplinary perspective*. Edited by C. Panter-Brick, R. Layton, and P. Rowley-Conwy, pp. 73–98. Cambridge: Cambridge University Press.

Trivers, R. L. 1971. The evolution of reciprocal altruism. *Quarterly Review of Biology* 46:35–57.

Tylor, E. B. 1924[1871]. *Primitive Culture*. 2 vols. 7th edition. New York: Brentano's.

Umapathy, G., M. Singh, and S. M. Mohnot. 2002. Status and distribution of *Macaca fascicularis umbrosa* in the Nicobar Islands, India. *International Journal of Primatology* 24:281–293.

Van de Velde, F. 1956. Rules for sharing the seals amongst the Arviligjuarmiut. *Eskimo* 41:3–6.

van der Sluys, C. M. I. 2000. "Gifts from the immortal ancestors: Cosmology and ideology of Jahai sharing," in *Hunters & gatherers in the modern world: Conflict, resistance, and self-determination*. Edited by P. P. Schweitzer, M. Biesele, and R. K. Hitchcock, pp. 427–454. Oxford: Berghahn Books.

van Schaik, C. P. 1989. "The ecology of social relationships amongst female primates," in *Comparative socioecology: The behavioural ecology of humans and other mammals*. Edited by V. Standen and R. A. Foley, pp. 195–218. Oxford: Blackwell.

van Schaik, C. P. 1999. The socioecology of fission-fusion sociality in orang-utans. *Primates* 40:69–86.

Vinicius, L. 2005. Human encephalization and developmental timing. *Journal of Human Evolution* 49:762–776.

Vokey, J. R., D. Rendall, J. M. Tangen, L. A. Parr, and F. B. M. de Waal. 2003. Visual kin recognition and family resemblance in chimpanzees (*Pan troglodytes*). *Journal of Comparative Psychology* 118:194–199.

Wakefield, M. L. 2008. Grouping patterns and competition among female *Pan troglodytes schweinfurthii* at Ngogo, Kibale National Park, Uganda. *International Journal of Primatology* 29:907–929.

Watters, D. E. 2006. *Notes on Kusunda grammar: A language isolate of Nepal*. Kathmandu: Himalayan linguistics. Archive 3.

Watts, D. P. 2002. What are friends for? The adaptive value of social bonds in primate groups. *Behaviour* 139:343–370.

West, G. B., J. H. Brown, and B. M. Enquist. 2001. A general model for ontogenetic growth. *Nature* 413:628–631.

Wheeler, P. E. 1992. The influence of the loss of functional body hair on the water budgets of early hominids. *Journal of Human Evolution* 23:379–388.

White, F. J. 1996. "Comparative socio-ecology of *Pan paniscus*," in *Great ape societies*. Edited by W. C. McGrew, L. F. Marchant, and T. Nishida, pp. 29–41. Cambridge: Cambridge University Press.

White, R. 1982. Rethinking the Middle/Upper Paleolithic transition. *Current Anthropology* 33:85–108.

Whiten, A., J. C. Goodale, W. C. McGrew, T. Nishida, V. Reynolds, Y. Sugiyama, C. E. G. Tutin, R. W. Wrangham, and C. Boesch. 1999. Cultures in chimpanzees. *Nature* 399:682–685.

Wich, S. A., E. H. M. Sterck, and S. S. Utami. 1999. Are orang-utan females as solitary as chimpanzee females? *Folia Primatologica* 70:23–28.

Wiessner, P. 2005. Norm enforcement among the Ju/'hoansi Bushmen. *Human Nature* 16:115–145.

———. 2009. Experimental games and games of life among the Ju/'hoan Bushmen. *Current Anthropology* 50:133–138.

Williams, G. C. 1966. *Adaptation and natural selection: A critique of some current evolutionary thought.* Princeton, NJ: Princeton University Press.

Wilson, D. S., and E. Sober. 1994. Reintroducing group selection to the human behavioral sciences. *Behavioral and Brain Sciences* 17:585–654.

Wilson, M. L., and R. W. Wrangham. 2003. Intergroup relations in chimpanzees. *Annual Review of Anthropology* 32:363–392.

Woodburn, J. 1982. Egalitarian societies. *Man (N.S.)* 17:431-451.

Wrangham, R. W. 1980. An ecological model of female-bonded primate groups. *Behaviour* 75:262–299.

———. 2000. "Why are male chimpanzees more gregarious than mothers? A scramble competition hypothesis," in *Primate males: Causes and consequences of variation in group composition.* Edited by P. M. Kappeler, pp. 248–258. Cambridge: Cambridge University Press.

Wrangham, R. W., J. Gittleman, and C. A. Chapman. 1993. Constraints on group size in primates and carnivores: Population density and day-range as assays of exploitation competition. *Behavioral Ecology and Sociobiology* 32:199–210.

Wrangham, R. W., and B. B. Smuts. 1980. Sex differences in the behavioural ecology of chimpanzees in the Gombe National Park, Tanzania. *Journal of Reproductive Fertility (Supplement)* 28:13–31.

Wynne-Edwards, V. C. 1962. *Animal dispersion in relation to social behavior.* Edinburgh, Scotland: Oliver & Boyd.

———. 1986. *Evolution through group selection.* Oxford: Blackwell Scientific Publications.

Yamakoshi, G. 2004. "Evolution of complex feeding techniques in primates: Is this the origin of great ape intelligence?" in *The evolution of thought: Evolutionary origins of great ape intelligence.* Edited by A. E. Russon and D. R. Begun, pp. 140–171. Cambridge: Cambridge University Press.

Yerkes, R. M. 1927. A program of anthropoid research. *American Journal of Psychology* 34:181–199.

Yoshiba, K. 1967. An ecological study of Hanuman langurs, *Presbytis entellus.* *Primates* 8:127–154.

Zhou, Q., F. Wei, C. Huang, M. Li, B. Ren, and B. Luo. 2007. Seasonal variation in the activity patterns and time budgets of *Trachypithecus francoisi* in the Nonggang Nature Reserve, China. *International Journal of Primatology* 28:657–671.

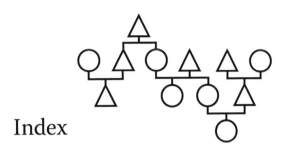

Index

Italicized page numbers indicate figures and tables.

Baldwin, James, 64–65, 126–27

behavior/s: action/reaction, 140; chimpanzees, and traditions versus rule-based, 119; cognition/behavior comparisons and, 20–21, 29–30, 95, 187; coordination of, 15–16, 166–69, 186; expression through, 21; grooming behavior variations and, 45–50, *46, 47, 48, 50,* 54n58, 100–104; normative behavior constraints and, 25, 128–29; Old World monkeys' dominance-submission, 37–38, *39,* 42–45, *43,* 48–49, 52, 54n6, 68; solitary, 139–40, 181nn1–2; variations in, 45–50, *46, 47, 48, 50,* 54n58, 100–105. *See also* chimpanzees; face-to-face interactions; grooming behavior variations; *Homo sapiens'* transition from chimpanzees; individualistic behavior; Old World (OW) monkeys; relational systems of social organization; social interactions; symmetric social interactions

Binford, Lewis, 170–72, 175

biological kin selection, and OW monkeys, 40–42, 51–52, 53nn2–3, 185, 191

biological kinship: boundaries based on, 66–68; cultural kinship and, 14, 66–68, 72–73, 179–80, 186, 191, 201n2; female philopatry for OW monkeys and, 40–41, 185, 191;

group-level properties and, 191, 201n2; hunter-gatherers and, 14, 66–68, 72–73, 179–80, 186. *See also* matrilineal units

bipedalism, 159. *See also Homo sapiens'* transition from chimpanzees

Birdsell, Joseph, 170, 172, 177

boundaries, 14, 68–73, 168–69, 186. *See also* social field size

Ceboids (New World monkeys), 35, 140, 141. *See also* Old World (OW) monkeys

Cercopithecoids (Old World monkeys). *See* Old World (OW) monkeys

Chapais, Bernard, 121–27

chimpanzees: overview, 9, 95–96, 134–35, 185; ancestor commonality with, 18–19, 33n1, 95–96; behavior traditions versus rule-based behavior and, 119; cognition and behavior comparisons and, 20–21, 187; community level behavior and, 99–104; cultural principles and, 18, *18*; dyads and, 104, 107, 113, *114,* 115, 143; face-to-face interactions and, 101, 111, 113, 116, 118, 121; food resources/social unit size relationship and, 102–3; grooming behavior variations and, 100–101; imitation transmission and, *118,* 118–19, 191–92, 201n1, 201n3; individual concepts

common chimpanzee (*Pan troglodytes* or Pt). *See Pan troglodytes* (common chimpanzee or Pt)

Fortes, Meyer, 15, 70, 129, 163, 179–80, 199–200

185; encephalization and, 36, 126–27, *150*, 159, 178; expression through behavior and, 21; female sex characteristics and, 161; imitation transmission and, *118*, 118–19, 131, 132, 201n1; individualistic behavior and, 9–10, 117–19; mate-guarding and, 122–25; mental representations distinctions and, 129–34, 135n4; normative behavior constraints and, 25, 128–29; pair-bonding and, 122–24, 127; scavenging activities and, 160; shared cultural meaning and, 131–32; social complexity and, 9–10, 117–19, *118*; social learning and, 117, 137, 190; social organization and, 25; STWM and, 13; working memory size and, 20–21, 149, *150*, 178. *See also* hunter-gatherers

hunter-gatherers: overview and description of, 9, 29–30, 55–57, 92, 183–84; ancestors' evolution to, 28–30; biological kinship versus cultural kinship and, 14, 66–68, 72–73, 179–80, 186; boundaries based on biological kinship and, 66–68; boundaries based on cultural kinship and, 68–73; collective ownership and, 52, 58, 83–87, 129, 184; cultural idea systems and, 167, 183–84; cultural kinship and, 30, 68–73, 137, 154–55, 184; culture as complex whole and, 65–66, 185–86; culture traits as analogue of genetic traits and, 64–65, 126–27; culture traits implementation and, 62–66; deviant individual control through social pressure and, 69–70, 91, 184, 194; egalitarianism and, 58, 87–91, 127–28, 184; heterogeneity of, 57–60, 93n2; hierarchical levels expansion and, 60–62; individual ownership and, 58, 84, 87, 129, 184, 190–96; IR versus DR and, 93n2; OW monkeys comparisons with, 29–30; ownership and sharing resources and, 58, 83–87, 90–91, 184; pair-bonding transition for, 122–24, 127, 159, 161, 181, 185; risk management with group-level selection and, 88–90, 194–96, *197,* 198–99; rules and, 25–26, 119, 183; self-identification as "real people" and, 29–30. *See also* culture; *Homo sapiens'* transition from chimpanzees; kin term maps; *specific groups*

imitation transmission: chimpanzees and, *118*, 118–19, 191–92, 201n1, 201n3; group-level properties and, *118*, 118–19, 191–92, 201n1, 201n3; *Homo sapiens'* transition from chimpanzees and, *118*, 118–19, 131, 132, 201n1

Immediate Return (IR), 93n2. *See also* hunter-gatherers

organization and, 98–99; performative patterns of social organization and, 104; reciprocal relations and, 151; social organization baseline pattern and, *39,* 39–40, 49. *See also* female philopatry; males/men

males/men: chimpanzees' transition from OW monkeys, and social units of, 101–4, 143; father categorization and, 158–61; man:woman/nature:culture analogy and, 86, 93n4; marriage and, 19, 21–26, 33n2, 59; mate-guarding and, 122–25, 135n3; parenting and, 158–61. *See also Homo sapiens'* transition from chimpanzees; male philopatry for OW monkeys

Mardu group, 70. *See also* Aborigines in Australia

marriage, 19, 21–26, 33n2, 59

mate-guarding, 122–25, 135n3. *See also Homo sapiens'* transition from chimpanzees

mating patterns changes (father categorization), 158–61. *See also Homo sapiens'* transition from chimpanzees

matrilineal units for OW monkeys: overview, 183, 191; agonistic encounters resolution and, 38, 42–43, 44–45, 46, 51–52, 54n7, 193; biological kin selection and, 40–42, 51–52, 53nn2–3, 191;

biological mothers/infants interactions and, 42–43, 53n5, 54n6; coalition formations and, 38, *39,* 41, 44–46, 51–52; cultural principles and, 19–21; dominance-submission behaviors and, 37–38, *39,* 42–45, *43,* 48–49, 52, 54n6, 68; grooming behavior variations and, 45–50, *46, 47, 48;* group size effects and, 109; individual ownership and, 193; ostensive forms of social organization and, 97–98, 105, 107; performative patterns of social organization and, 97–98, 107; social organization/social complexity and, 114–16; traits implementation and, 62–63, 68. *See also* female philopatry for OW monkeys; Old World (OW) monkeys

men/males. *See* male philopatry for OW monkeys; males/men

mental representations, 129–34, 135n4. *See also Homo sapiens'* transition from chimpanzees

mind, theory of. *See* theory of mind

Neanderthals, 28–30, 189

neocortex ratio: group size and, 49, *50,* 54n8, 107–10, *108, 110,* 135n1, 178; social complexity and, 108–9, 113–17, *114, 115,* 178; social organization and, 114–17, *115. See also* Old World (OW) monkeys

Netsilik Inuit, 58–59, 68–69, 82–83, 85–88, 90–91, 193–96. *See also* Inuit

New World monkeys (*Ceboids*), 35, 140, 141. *See also* Old World (OW) monkeys

Ngogo, Uganda, chimpanzees, 100–101, 104, 192

Nissen, H. W., 111, 135n2

normative behavior constraints, 25, 128–29. *See also Homo sapiens'* transition from chimpanzees

Old World (OW) monkeys: overview, 9, 35–36, *37,* 50–52, 183; ancestor commonality with, 36, *37;* community level variation and, 99–100; cultural principles and, 19–21; dominance hierarchies and, 37, 38, 63, *106;* dyads and, 40–41, 53n1; face-to-face interactions and, 52, 58, 178; hunter-gather groups comparison with, 29–30; infanticide and, 37, *38,* 42, 53n4; ostensive forms of social organization and, 105, 107; reciprocal altruism and, 51, 187–88; social interactions and, 140–41; social learning and, 117, 190, 201n1; social organization as emergent form and, 42–45, *43, 46,* 53nn4–5, 54nn6–7, 54n7; social organization baseline pattern and, 36–40, *38, 39;* solitary behavior and, 139–40, 181nn1–2; species level

variation and, *38,* 99; traits implementation and, 62–65, 68. *See also* chimpanzees; female philopatry for OW monkeys; grooming behavior variations; male philopatry for OW monkeys; matrilineal units for OW monkeys; neocortex ratio

Ortner, Sherry, 93n4

ostensive forms of social organization, 97–98, 105, 107. *See also* matrilineal units for OW monkeys

OW (Old World) monkeys. *See* Old World (OW) monkeys

ownership: collective, 52, 58, 83–87, 129, 184; individual, 58, 84, 87, 129, 184; sharing resources and, 58, 83–87, 90–91, 93nn3–4, 154, 184. *See also* group-level properties; hunter-gatherers

pair-bonding, 23, 98, 122–24, 127, 159, 161, 181, 185. *See also* hunter-gatherers

Pan paniscus (pygmy chimpanzee or Pp): behavior variations and, 105; cognition/behavior comparisons and, 20, 95; dominance hierarchies for male social units and, 102; food resources/social unit size relationship and, 102–3; mate-guarding and, 135n3; reciprocal relations and, 151; resource sharing and, 160–61; social brain hypothesis/social complexity and, *50*, 108, *108*, 109; social organization variation and, *38*, 99–100

Pan troglodytes (common chimpanzee or Pt): asocial behavior and, 140; behavior variations and, 105; cognition/behavior comparisons and, 20, 95; dominance hierarchies for male social units and, 102; food resources/social unit size relationship and, 102–3; mate-guarding and, 135n3; reciprocal relations and, 151; resource sharing and, 160–61; social brain hypothesis/social complexity and, *50*, 108, *108*, 109; social organization variation and, *38*, 98–100

parenting: male, 158–61; Old World monkeys' biological mothers/infants interactions and, 42–43, 53n5, 54n6. *See also* females/women; *Homo sapiens*' transition from chimpanzees; males/men; matrilineal units for OW monkeys

Parson, Talcott, 141

pastoral societies, 55–56, 60, 93n1

performative patterns of social organization, 97–98, 103–4, 105, 107

philopatry, 39, 40, 53n1. *See also* female philopatry for OW monkeys; male philopatry for OW monkeys

prescriptive altruism (Axiom of Amity), 15, 70, 129, 163, 179–80, 199–200. *See also* altruism

pygmy chimpanzee (*Pan paniscus* or Pp). *See Pan paniscus* (pygmy chimpanzee or Pp)

Radcliffe-Brown, Alfred, 29, 72, 80, *81*

"real people," self-identification as, 29–30, 68, 173

reciprocal altruism, 51, 90–91, 179–80, 187–88. *See also* Axiom of Amity (prescriptive altruism); reciprocal relations

reciprocal relations: cognition and symmetric social interactions and, 149–52, 179, 182n6; female and male philopatry for OW monkeys and, 151; genealogical kinship transition and, 156–57, *157*, 180–81; theory of mind and, 156–57, *157*, 163–65, *166*, 180–81, 182n7. *See also* reciprocal altruism

recursive composition of relations, 155–56, 179, 187

recursive computation of relations, 148–49, *150*, 182n5, 187

relational systems of social organization: overview, 10, 14–15, 137–38, 178–81; action/reaction behavior and, 140; asocial behavior and, 138–40, 181n1; behavior coordination and, 15–16, 166–69,

social field size: overview, 169–70, 177, 181, 188, 200; data heterogeneity and, 171–72, *172*; date and prediction agreement and, 174–77, *175, 177*; empirically based modal value/population size and, 170–71; modal value prediction/population size and, 173–74; population size constraints/cultural kinship and, 173, 186, 188, 191, 201n2. *See also* boundaries; relational systems of social organization

social interactions: overview, 140–42, 182n3; dyads and, 141–42, 182n3; functionality of projected relation and, 162–63, *164*; Old World monkeys and, 140–41; symmetric social interactions and, 142, 182n3. *See also* symmetric social interactions; theory of mind

social learning: chimpanzees and, 118, *118*, 134, 135, 190, 201n1; *Homo sapiens'* transition from chimpanzees and, 117, 137, 190; OW monkeys and, 117, 190, 201n1. *See also* culture; social organization

social organization: overview and uniqueness of, 11–12, 15–18, 32, 199–200; ancestor commonality/disconnect and, 18–19, 24–26, 33n1, 95–96; ancestors' evolution to hunter-gatherers and, 28–29; biological evolution discontinuity/continuity and, 26–27; chimpanzees, and rudiments

of, 18, *18*, 18–19, 20–21, 33n1; cognition/behavior comparisons and, 20–21, 29–30, 95, 187; community level behavior for chimpanzees and, 99–104; as emergent form, 42–45, *43, 46*, 53nn4–5, 54nn6–7, 54n7; evolutionary change and, 31–32, 188–90; food resources/social unit size relationship and, 102–3; genetic trait co-evolution and, 56–57, 93n1; hunter-gatherers and, 28–30; kin term maps, and structure for, 80–83; lesser and greater apes, and variation in, 98–99, 185; lesser and greater apes' variation in, 98–99; neocortex ratio and, 114–17, *115*; ostensive forms of social organization and, 97–98, 105, 107; OW monkey baseline patterns and, 36–40, *38, 39*, 49; OW monkey matrilineal units as rudiments of, 19–20, 21; performative patterns for chimpanzees and, 97–98, 103–4, 105, 107; performative patterns of, 97–98, 103–4, 105, 107; relational systems of social organization and, 138–44; rudiments of, 17–21, *18*; social complexity and, 60–62, *108*, 114–17, *115*, 186; STWM and, 12–14; variation in chimpanzees' and, 99–104. *See also* cultural kinship (kinship terminology); culture; hunter-gatherers; relational systems of social organization; social learning

About the Author

Dwight W. Read is a Distinguished Professor at the University of California, with appointments in the Department of Anthropology and Department of Statistics at UCLA. He is a recognized expert in kinship theory and cultural evolution, among other research areas. His authoritative book, *Artifact Classification: A Theoretical and Methodological Approach,* was published by Left Coast Press Inc. in 2007. Dr. Read is currently completing another new book with Murray Leaf, *The Conceptual Foundation of Human Society and Thought: Anthropology on a New Plane.* He has written over 150 articles on anthropological theory, archaeological analytic methods, kinship, computer applications, and biological anthropology, and has been involved in field projects from California to the Kalahari Desert in Southern Africa.